Spooky

Ghost, Ghouls and Top 40 Radio

Larry Wilson

Chiller Books

ACKNOWLEDGEMENTS

To my family, friends, and colleagues—thank you for your unwavering patience and support during the writing of this book. Your understanding and encouragement have been my foundation through late nights and long hours spent delving into the unknown.

A special thanks to those who opened their homes, businesses, and properties to me, allowing me the rare opportunity to experience firsthand the strangeness that defies explanation. Without your trust, these investigations would not have been possible. Your willingness to share your spaces and your stories has enriched this journey, and for that, I am truly grateful.

Contents

DEDICATION

This book is dedicated to my wife, Kathy, and my son, Cory—your love and support have been my anchor throughout this journey into the unknown.

To all who have felt the unsettling brush of the supernatural, who have stood face-to-face with the unexplained, and to those who continue to seek understanding in the midst of strangeness—this is for you.

And to my fellow investigators, those who tirelessly pursue answers to the extraordinary and elusive questions the paranormal presents—may we continue to push the boundaries of what we think we know.

All Honor and Glory to God the Father Almighty for ever and ever.

Spooky

Ghost, Ghouls and Top 40 Radio

LARRY WILSON

INTRODUCTION

When I embarked on my journey as a paranormal investigator over two decades ago, I had no inkling of the unexplainable phenomena that awaited me. What began as a simple curiosity soon evolved into a full-blown obsession, fueled by the strange and extraordinary encounters I experienced. My initial goal was to determine whether the supernatural was real or merely the product of overactive imaginations. I sought to capture definitive evidence through the use of advanced audio and video technology.

However, as time passed, I came to a sobering realization: in the realm of the paranormal, there is no such thing as conclusive evidence. Despite the capabilities of modern technology, recordings of odd and extraordinary events often left more questions than answers. The reality is that one must witness and experience these unbelievable occurrences firsthand to truly comprehend their authenticity.

Throughout my adventures in the world of the unexplained, I have met a myriad of fascinating individuals and forged lasting friendships. One such friend is Jason Bond, known to his listeners as "Bondsy," the host of the popular morning radio show *The Morning Grind* on 99.7 THE MIX in Springfield, Illinois.

In 2011, I published my first book on the paranormal, *Chasing Shadows*, which chronicled my investigations and experiences across the Midwest. The book's release led to my introduction to

Bondsy in the fall of that year. He invited me to join his annual Halloween show to discuss the paranormal and share some of my most intriguing encounters. While Bondsy didn't outright dismiss my stories, his natural skepticism drove him to seek the truth for himself.

Intrigued by the unknown, Bondsy accepted my invitation to join me on a few investigations. This partnership blossomed into an annual tradition: each year since 2011, Bondsy and I have conducted a paranormal investigation or two, culminating in a live Halloween morning broadcast. During the show, we delve into the details of our investigations, sharing audio evidence of unexplained voices and sounds captured during our explorations.

Through these experiences, we have not only entertained and intrigued our audience but also deepened our own understanding of the mysterious world that exists just beyond the veil of the ordinary.

This book documents the unusual and sometimes mind-boggling phenomena that Bondsy and I, along with a dedicated team of interns and staff from the radio station, have witnessed and encountered over the years. Our journey has taken us to the brink of the unknown, where the ordinary collides with the extraordinary and the boundaries of reality blur.

As you delve into these pages, I invite you to lock your doors and immerse yourself in the stories that have both baffled and intrigued us. Pour yourself a cup of your favorite coffee or fill your glass with a comforting drink and prepare to embark on an incredible journey into the unknown.

What you will read is not just a collection of eerie tales and strange occurrences but a testament to the human spirit's relentless pursuit of truth in the face of the inexplicable. Each chapter reveals the meticulous investigations and tireless efforts to understand the mysteries that lie beyond our everyday experience. We have documented voices from the void, shadows that dance in the night, and encounters that defy conventional explanation.

Through it all, our quest has been driven by a deep-seated curiosity and a desire to peel back the layers of the unknown. We have faced skepticism and doubt, but each investigation has only strengthened our resolve. This book is a chronicle of that journey—a journey that continues to challenge our perceptions and ignite our imaginations.

So, sit back and let the stories unfold. Allow yourself to be drawn into a world where the line between reality and the paranormal is thin, and where every investigation opens up new possibilities. This is not just a book; it is an invitation to explore the mysteries that surround us and to join us in our ongoing quest to understand the unexplainable.

Prepare yourself for an incredible expedition into the unknown. The truth, as they say, is out there—and it is far stranger than we could ever have imagined.

RIDGE CEMETERY
SHELBY COUNTY, ILLINOIS

ONE

The Graveyard

I first met Bondsy in October of 2011, back when the Springfield, Illinois, radio station was known as 99.7 KISS FM. He invited me to be a guest on his morning show to discuss the paranormal and share some of my most compelling experiences as an investigator of the unknown. The studio was abuzz with the usual morning chaos, but there was an undercurrent of excitement about the unusual topic at hand.

Bondsy, the charismatic host with a voice that could both soothe and invigorate, greeted me warmly. His genuine curiosity about the paranormal was evident, but so was his skepticism. As I recounted tales of ghostly apparitions, unexplained sounds, and eerie encounters, I could see the wheels turning in his mind. He was intrigued, yet cautious—wary of accepting the phenomena without tangible proof.

After sharing several stories, Bondsy leaned back in his chair, a thoughtful expression on his face. It wasn't that he outright dismissed my accounts, but his natural skepticism demanded a firsthand experience. He wanted to see if the supernatural was real for himself, to venture into the unknown and come back with his own conclusions.

Recognizing his genuine interest, I extended an invitation for him to join me on a few investigations. I promised him an insider's look at the meticulous process of a paranormal investigator, from setting up equipment to analyzing the findings. It was an offer he couldn't resist. As they say, the rest is history.

From that moment on, our partnership took us to some of the most haunted and mysterious locations in the Midwest. Bondsy's transformation from a skeptic to a curious explorer of the paranormal was remarkable. Each investigation not only brought us closer as colleagues but also deepened our understanding of the obscure world we were probing.

Thus began an adventure that neither of us could have predicted.

What started as a simple radio show appearance evolved into a shared journey into the unknown, where every investigation uncovered new mysteries, and every story added another piece to the puzzle. Together, we ventured into the heart of the paranormal, driven by a shared desire to uncover the truth and challenge the boundaries of what we thought we knew about the world around us.

Since my 2011 radio appearance, the station has made it a tradition to broadcast a live show every Halloween morning. It became an eagerly anticipated event, where listeners tuned in to hear tales of the paranormal and the latest findings from our investigations. By the time the 2012 show was approaching, Bondsy had an idea that would elevate the experience for our audience. He proposed that we investigate prior to Halloween, pre-record it, and then provide a live on-air recap on Halloween morning.

Eager to take on this new challenge, Bondsy turned to me and asked if I knew of a location that would make for an unforgettable investigation. Without hesitation, I mentioned Shelby County, Illinois's Ridge Cemetery, located at Williamsburg Hill. This site, shrouded in mystery and known for its intense paranormal activity, was a place where I had encountered inexplicable phenomena on several occasions.

I described to Bondsy the eerie atmosphere of Ridge Cemetery, a place where the veil between the living and the dead seemed remarkably thin. The cemetery, perched atop an 800-foot-tall hill and surrounded by dense woods, had a reputation for strange occurrences. For decades, bizarre events have occurred at this secluded Illinois graveyard that range from perplexing to terrifying. Ghostly figures, mutilated cattle, UFOs, and orbs of light with dazzling maneuverability have been witnessed. Ghoulish hooded figures in black, a phantom old man, and a woman in black who vanish have also been seen by people I've interviewed. Several times, I've heard disembodied voices and muffled screams coming from beneath the ground.

I recounted some of my personal experiences there—nights when the air felt charged with energy, and moments when I heard voices and sounds that defied logical explanation. The site was a hotspot for paranormal activity, making it the perfect backdrop for our Halloween special.

Bondsy, intrigued and slightly unnerved by my descriptions, agreed that Ridge Cemetery would be the ideal location for our pre-Halloween investigation. We began planning the logistics, gathering our equipment, and preparing ourselves for whatever we might encounter. The stage was set for what promised to be an unforgettable night.

As the date of the investigation approached, anticipation built. We knew we were about to embark on a journey that would not only test our nerves but also push the boundaries of our understanding of the paranormal. This was more than just a show—it was a chance to delve deeper into the mysteries that had captivated us both and to share our findings with an audience eager to explore the unknown.

Little did we know the events that would unfold at Ridge Cemetery would be more extraordinary than anything we had anticipated, setting the tone for an annual tradition that would continue to captivate and mystify our listeners for years to come.

Several days before our adventure, I called Bondsy to coordinate the details of our upcoming investigation. We needed to establish who would be joining us, as well as the logistics of when and where I would pick them up. After a brief discussion, we decided that his co-host, Sarah, and intern Taylor Fishburn, who went by the quirky on-air nickname "Tuna-melt" or simply "Tuna," would accompany us on our ghost-hunting expedition.

Sarah was known for her quick wit and skeptical demeanor, which promised to add an interesting dynamic to our team. Tuna, on the other hand, was eager and enthusiastic—a perfect balance to the group's composition. Despite the lightheartedness of her nickname, Tuna's presence would prove to be unexpectedly significant, as you will come to understand as our story unfolds.

As we ironed out the final details, a sense of anticipation began to build. Hopefully, our team was ready to confront whatever mysteries the night might hold.

The plan was for me to pick up Sarah and Taylor at the station. Bondsy would be running late because of a family commitment, so he would meet us at the local JCPenney parking lot in Springfield, and we would leave from there to make the fifty-mile trek to Williamsburg Hill.

While coordinating our plans, I mentioned to Bondsy that I wanted to conduct an experiment with Sarah and Tuna. I told him about a particular grave—a young boy's resting place—where women, for reasons unknown, become inexplicably emotional when near it. The phenomenon was both intriguing and baffling, and it demanded a controlled approach to verify its authenticity.

I explained to Bondsy that the experiment's success hinged on the girls not knowing what to expect. From past experience, I had observed that the grave seemed to affect only women who were unaware of its peculiar history—their reactions pure and unprompted. This spontaneous emotional response was a critical component of the phenomenon, one that could not be replicated if they were pre-warned or anticipating something unusual. Bondsy listened intently, his curiosity piqued. He understood the importance of maintaining the integrity of the experiment.

Sarah, despite having accompanied me on two prior investigations after my initial appearance on the radio station, was still relatively new to the world of the paranormal. Tuna, on the other hand, was stepping into uncharted territory. This would be Bondsy and Tuna's first real brush with the supernatural, and I wanted it to be as authentic and impactful as possible.

As we finalized our plans, I could sense Bondsy's excitement mixed with a hint of apprehension. The prospect of witnessing something inexplicable was both thrilling and daunting. For Sarah and Tuna, the night promised to be an eye-opening introduction to the mysteries that lay beyond the veil of the ordinary.

With everything set, I hung up the phone and began preparing our equipment. Cameras, audio recorders, and EMF detectors were meticulously checked and packed. Each piece of gear had a vital role in capturing evidence of the paranormal phenomena we hoped to encounter.

When I picked up Sarah and Tuna at the station, Sarah suggested I share a few stories about my paranormal encounters over the years. I was hesitant, knowing that the tales could be unsettling for someone unaccustomed to the strange and unexplained. But Sarah insisted, and so, as we drove through the quiet streets heading to our meeting place with Bondsy, I recounted some of the more memorable moments from my investigations.

With each story I told, I could see Tuna's nerves wearing thin. It was clear that the weight of my experiences was starting to bear down on her. By the time we arrived at our rendezvous point, Tuna was visibly anxious, her usual bubbly demeanor replaced by a tense silence.

Bondsy arrived promptly at 9:00 p.m., and we all piled into the car, setting off toward Williamsburg Hill for our paranormal adventure. The drive was filled with a mix of excitement and apprehension. We made a brief stop at a convenience store in Tower Hill to pick up drinks and snacks—a final moment of normalcy before we plunged into the unknown.

I turned onto Road 1100 East and followed the winding path to the top of Williamsburg Hill. As we neared the graveyard, Tuna's anxiety became palpable. By day, Williamsburg Hill is a picturesque spot, but under the cover of night, it transforms into a place of foreboding shadows and eerie stillness.

At the top of the hill, we made a left at the tall microwave tower and proceeded down the narrow road leading to the graveyard. The atmosphere grew more oppressive with each passing moment. Suddenly, Tuna's voice, tight with fear, cut through the silence. "Stop the car, I'm going to get sick!"

Her fear was so intense that she was physically ill. The situation wasn't helped by the fact that my CD player was eerily playing the theme song from the movie Halloween. Sensing her genuine distress, I muted the music and pulled over. We waited for a couple of minutes, allowing Tuna to regain her composure. Once her nausea subsided, we continued our journey down the gravel road, which ended at the cemetery gate.

My headlights illuminated the spooky graveyard under the pallid moon. A frigid wind rustled the brown leaves, creating an atmosphere that seemed perfect for ghouls, goblins, and ghosts. As the road dead-ended, the cemetery seemed to appear out of nowhere, exuding an eerie sensation that someone or something was always waiting for its next visitor.

On overcast nights or during a new moon, Ridge Cemetery is one of the darkest places I have ever investigated. Its isolation and the heavy shadows that engulf it make it a prime location for paranormal activity. Over the years, I have spent many nights out here alone, and each time, without fail, I felt watched and followed. The sensation of unseen eyes tracking your every move is unnerving, to say the least.

For those seeking a genuine thrill, Ridge Cemetery offers an experience like no other. Venture out here alone some night, and you won't be disappointed. The darkness, the stillness, and the sense of being watched will provide an adrenaline rush that lingers long after you leave. Tonight, with Bondsy, Sarah, and Tuna by my side, we were about to uncover the mysteries hidden within this eerie resting place, stepping deeper into the unknown with each cautious step.

As we pulled in, I made a U-turn and parked so that my vehicle was facing the direction I had just come from. I always do this but am not sure why. Maybe it's easier to unload equipment, or maybe it's in case I see something so terrifying I need to make a quick getaway.

As we exited my vehicle, the darkness of the graveyard immediately surrounded us. The night was so black it seemed to swallow the beams of our flashlights. Standing there, adjusting to the inky void, one of the girls broke the silence. "I can't believe how dark it is," she said, her voice tinged with a mix of awe and trepidation. The cemetery's ancient tombstones, barely visible in the dim light, added to the sense of unease.

For the investigation, we would use digital audio recorders and a video camera. But the most important piece of equipment on this night would turn out to be the SB-7 Spirit Box. The SB-7 Spirit Box is a popular tool used in the field of paranormal investigations to facilitate potential communication with spirits by generating white noise through rapidly scanning either Am or FM radio frequencies. Proper usage involves careful preparation, clear questioning, and thorough analysis of recorded sessions to identify possible paranormal responses. In a bit you will read how the SB-7 device left us all shaking our heads in disbelief.

We carried our equipment to the southeast corner of the graveyard and set up our base camp for the night. Over the years, I've discovered that this particular area is a hotspot for paranormal activity. In conducting more than a hundred investigations at Ridge Cemetery, the southeast corner has consistently yielded the most compelling evidence.

As we arranged our gear, a sense of anticipation filled the air. Although we didn't realize it at the time, we were mere moments away from capturing our first disembodied voice of the night.

Shortly after we set up base camp, Bondsy began recording a segment for the Halloween show using a recorder from the station. Sarah and I were nearby, setting up additional equipment, while Tuna-melt was exploring a nearby grave. The night was still, the only sounds being our hushed conversations, the occasional rustle of leaves and the distant call of coyote.

It wasn't until a few days later, during a thorough review of the recordings, that we discovered the chilling evidence. As Bondsy spoke into the recorder, a voice—unmistakably not belonging to any of us, somehow embedded itself in his recording. It was clear and distinct, yet eerily disembodied, as if it had materialized out of thin air.

Before I describe what happened, I should explain that the relationship between Bondsy and Sarah is more like that of a brother and sister than a professional one. They are good friends and, just like siblings, they often have good-natured arguments. This was the case when the EVP was recorded.

As Bondsy was recording the sound bite for the show, he looked at Sarah and asked, "Have you guys seen anything yet?"

Sarah, who was more unnerved by the thought of coyotes or possibly drug dealers showing up in the secluded location than she was about ghosts or spirits, gave Bondsy a worried look but didn't answer him.

A bit perturbed that she didn't respond, Bondsy said, "Okay, you two city broads need to calm down," referring to Sarah and Tuna.

Sarah fired back, "Okay, Mr. St. Louis," insinuating that Bondsy, who had grown up in St. Louis, Missouri, was no outdoorsman either.

Bondsy replied, "Where do I live?"

To which Sarah answered, "You live on a numbered street!"

Unbeknownst to us at the time, as soon as Sarah replied, "You live on a numbered street," a voice came out of nowhere and said, "I feel Sarah."

It wasn't until several days later, when reviewing our audio recordings, that we discovered the disembodied voice.

When I heard the voice, it was unmistakable—loud, clear, and disturbingly close. It sounded like an adolescent male, speaking with a clarity that sent a chill down my spine. The recorder, cradled in Bondsy's hand, captured every nuance of the voice as if the speaker were standing right beside us. If they were that close to the recorder, they were undoubtedly close to us as well, which was an unsettling realization.

But it wasn't just the proximity that unnerved me. The voice had uttered Sarah's name, as clear as day. This wasn't some random auditory anomaly, or a fleeting sound wave caught by chance. It was deliberate, intelligent, and personal. Whoever—or whatever—was with us that night had not only been close by but had also been listening. Listening to our conversations, absorbing our words, and then choosing the perfect moment to make its presence known.

For me, this was confirmation beyond a shadow of a doubt. We weren't dealing with residual energy or a mere echo from the past. This was something conscious, something that understood who we were and was actively engaging with us. It knew Sarah's name.

In the world of paranormal investigation, moments like this are rare but profound. They cut through the fog of skepticism and force you to confront the reality that there are forces out there we don't fully understand—forces that can observe us, interact with us, and even communicate with us on a level that defies logical explanation.

After listening to the recording, I couldn't shake the feeling that we had just crossed a threshold. The voice wasn't just a piece of evidence; it was a message, a signal from the other side that we were not alone in that dark, deserted cemetery.

We didn't hear the voice in real-time, so after Bondsy and Sarah's brief exchange, Sarah and I walked back to my vehicle to get my reading glasses, which I had left on the front seat. When we returned, I decided it would be a good time to conduct our first SB-7 Spirit Box session of the night. I had used the box many times before but had never used it in a graveyard, so I wasn't sure what to expect.

We set up our session at base camp, close to the gravesite of the young man whose story had drawn us there. The atmosphere was thick with anticipation—the kind that settles in when you know something extraordinary is about to happen. Sarah and Tuna-melt were seated in camping chairs, their faces illuminated by the soft glow of their flashlights. I stood nearby, keeping a watchful eye on the surroundings, while Bondsy, ever the skeptic soon to be believer, had chosen a wooden bench at the edge of the cemetery as his vantage point.

The bench sat at the boundary where the cemetery met the dense timber, a line that seemed to separate the living from whatever spirits might be lurking in the shadows. Bondsy had a reason for his choice, a reason that added an extra layer of tension to the night. Earlier in the evening, I had recounted a story from the summer of 2010. An investigator named Chris had once sat on that very bench, casually observing the area. But his calm demeanor quickly shattered when he leaped from the bench, his face pale and his breath ragged. He swore he felt someone standing behind him—an unseen presence that sent a jolt of fear through him like electricity.

As Bondsy settled onto the bench, his back to the timber, I could tell he was bracing himself for whatever might come. He wasn't just sitting there; he was testing the boundaries, daring the unknown to make its move. The rest of us fell into a tense silence, the weight of the story lingering in the cool night air.

The bench itself seemed to carry a charge, a history that resonated with the experiences of those who had dared to sit there before. Bondsy's decision to place himself in that same position was a testament to his curiosity, but also to his growing respect for the forces we were dealing with. This wasn't just about capturing evidence anymore; it was about understanding, about feeling the presence of something that defied the rules of our world.

As the session began, I kept one eye on the equipment and the other on Bondsy, wondering if he would feel the same unsettling sensation that Chris had described. The timber behind him loomed large, its darkness impenetrable, as if it were holding its breath, waiting for the right moment to reveal what lay within.

Sarah was the only one from the radio station who had seen the spirit box in action before. She knew what to expect from this obscure device, having witnessed its eerie capabilities firsthand. Bondsy, on the other hand, had only heard me talk about it in passing—how strange, disembodied voices had come through the static, seemingly in response to our questions. But being the skeptic he was, he needed to see it with his own eyes, hear it with his own ears, before he would even entertain the possibility that it worked.

For those unfamiliar with a spirit box, it's an unsettling piece of equipment. Most of the time, you hear nothing but loud static or white noise—the machine rapidly scans through radio frequencies at a rate so fast that no station can be clearly heard. Occasionally, though, a word or phrase cuts through the noise, defying the logic of how the device should operate. Given the speed at which it scans, full words, let alone coherent statements, shouldn't have time to form. Yet, they do.

Over the years, I have used the box in countless investigations, asking questions tailored to each location, and I have received responses that were startlingly relevant. These weren't just random radio snippets; they were answers that seemed directly connected to the questions I posed. Once, in 2014, I asked, "Who is the president of the United States?" Almost instantly, three distinct voices, each coming through on different frequencies and in different tones— male, female, young, and old—responded with the same answer: "Obama!" The synchronization and clarity left no room for doubt.

Another time, just as an investigator named Chris turned the box on, a voice came through, clear and accusatory: "Shame on Chris!" It wasn't just the words that were unsettling; it was the way they emerged, as if spoken by the same entity but on multiple frequencies. This wasn't a random piece of radio chatter; this was something else—something that knew who Chris was and had a message to deliver.

As I explained all this to Bondsy, I could see the wheels turning in his mind. He was intrigued, maybe even a bit unnerved, but still holding onto his skepticism. The spirit box, with its mix of technology and the inexplicable, challenges even the most rational minds. In a matter of minutes, Bondsy would experience that challenge for himself.

We powered up the device, and the static roared to life, filling the night air with its relentless hum. The atmosphere around us seemed to thicken as we waited, every crackle of the box a potential message from the other side. This was no longer just an experiment; it was a confrontation with the unknown, and we were all about to see where it would lead.

If I had been asked to pick one word that would shatter any lingering doubts Bondsy, Sarah, and Taylor had about the spirit box, I couldn't have orchestrated a better moment than what happened that night during the KISS FM investigation.

We were huddled around the device, the air thick with anticipation, and the grating sound of static filled the silence. It was the kind of noise that can wear down your nerves—a relentless barrage of white noise that usually masks anything of significance. But then, cutting through that static, as clear as a bell, a voice rang out, loud and unmistakable: "Tuna!"

Taylor's head snapped up, her eyes wide with shock. "Did that just say my name?" she asked, her voice trembling slightly. Sarah's reaction was instant and incredulous: "What!" she blurted out, while Bondsy, ever the skeptic, let loose with a more colorful choice of words. We all heard it—there was no mistaking it. The voice had said "Tuna," Taylor's on-air nickname.

The voice was male, and its clarity was both remarkable and unsettling. In that instant, any remaining skepticism in the group started to crumble. The spirit box had spoken a name that wasn't just relevant—it was specific, personal, and impossible to dismiss as mere coincidence.

And yet, as unbelievable as this moment was, it was only the beginning. Throughout the night, during every box session we conducted, that same voice came through again and again, repeating "Tuna" with an eerie consistency. Sometimes it was said once, sometimes multiple times, but it was always clear and always directed at Taylor.

Each time the voice came through, the tension in the air ratcheted up another notch. It was as if the entity—whatever it was—had fixated on Taylor, choosing her as the focal point for its communication. The experience was beyond anything we could have anticipated, and it left us all shaken—not just by the reality of what we were encountering, but by the undeniable intelligence behind it.

That night, the spirit box didn't just deliver random words or fragmented phrases; it delivered a message—one that resonated deeply with all of us and confirmed that we were dealing with something far beyond the ordinary.

A comment Bondsy made that night served as a clear sign that he finally believed the spirit box was more than just a gadget. It happened after we heard "Tuna" come through the box several more times. He stood there, shaking his head in disbelief, trying to wrap his mind around what was happening.

"Man, this makes no sense!" he finally blurted out, his voice tinged with equal parts wonder and frustration. "I mean, hearing the name 'Tuna,' which is the nickname of one of the people with us, coming through a radio box that's scanning multiple frequencies— in the middle of the night, in a cemetery, located out in the middle of nowhere—doesn't make sense! Unless someone is out here on an amateur radio doing a freaking Chicken of the Sea commercial for Tuna, you shouldn't get that word coming through a radio box!"

Bondsy's offhand comment, laced with humor, was really an admission that what we were experiencing defied logic. There was no rational explanation for it—no one was out there in the dark broadcasting commercials or playing pranks. We were alone, surrounded by nothing but tombstones and the whispering wind.

So, we could pretty much rule out the possibility of a rogue radio operator in the middle of Williamsburg Hill, pushing a canned tuna agenda. What we were left with was something far stranger—an encounter that not only challenged Bondsy's skepticism but also left all of us questioning the very nature of what we were dealing with. The spirit box had delivered a message that night, and whether we liked it or not, it was impossible to ignore.

We heard "Tuna" come through the box multiple times that night, and curiosity got the better of me. I turned to Taylor and asked if she'd be willing to hold the box herself as a sort of experiment. She agreed, so I handed it over to her. The moment it was in her hands, a clear male voice cut through the static: "Tuna!" We all stood there, stunned. It was as if the box—or whatever was communicating through it—had been waiting for this moment.

It was hard to wrap our minds around what was happening. Why was this voice so fixated on Taylor? It didn't make any logical sense, but it was impossible to ignore. I decided to push further, to see if we could establish a personal connection. I asked Taylor if anyone close to her had passed away recently. She paused for a moment, then told us about her grandfather on her mother's side. He had passed not long ago, and she had been very close to him. The way she spoke of him, you could tell he meant a lot to her.

She mentioned that her grandfather had a nickname—Dandy. The gears in my mind started turning. What if this wasn't just some random spirit, but someone with a direct connection to Taylor? I suggested she ask a question that only her grandfather would know the answer to. Taylor thought about it, then spoke into the static: "Grandpa Dandy, this is Taylor. Can you tell me what color my hair was when I was a little girl?"

We held our breath, waiting for what felt like an eternity. Then, cutting through the white noise, came a single word: "Red." The voice was male, confident, and unmistakable.

Taylor's eyes widened in shock. "No way," she gasped. "Until I was seven years old, I had red hair! Grandpa Dandy would definitely have known that!"

At that moment, the atmosphere around us shifted. What had started as an experiment was quickly becoming something far more profound. We weren't just picking up random voices; we were communicating with something—or someone—that knew intimate details about Taylor's life. It was as if a bridge had been built between our world and a world beyond, and Taylor's grandfather was standing on the other side, reaching out to his granddaughter in a way that was as startling as it was comforting. There was no denying that something very real—and very personal—was happening right before our eyes.

From the look on Taylor's face and the tremor in her voice, it was clear she was beginning to believe that, just maybe, this was indeed Grand-Pa Dandy trying to reach out to her. She asked several more questions, her voice wavering between skepticism and hope, but while we could hear responses, they were garbled, like whispers carried off by the wind before they could fully form. The connection, it seemed, was fading.

Then, just as suddenly as it began, the voices ceased altogether. An unsettling silence settled over us. I could sense the disappointment creeping in, and I decided it was time to give the box a rest. But before doing so, I suggested Taylor give her grandfather one last chance to speak. "Ask him if he has anything else to say before we turn the box off," I urged.

Taylor nodded, took a deep breath, and called out into the static, "Grand-Pa Dandy, we're going to turn the box off now. Is there anything else you want to say?"

We waited, the night pressing in around us, our ears straining to catch any response. Then, cutting through the thick static like a knife, came a single word: "Stay."

The voice was unmistakable—male, clear, and deliberate. It wasn't a plea or a demand, but more of a gentle request. Was this truly Tuna's grandfather, asking her to stay a little longer, to keep the connection alive for just a few more moments? The weight of the question hung in the air, heavy and unanswerable. There was no way to know for sure.

In that moment, we were left with more questions than answers. The paranormal is often like that—a tangle of mysteries, where even the most compelling evidence leaves you grasping for something more. All we knew for certain was that something—or someone—had reached out to us from beyond, leaving us with an experience that would linger long after we left that haunted hilltop.

When we wrapped up the SB7 box session, I decided it was time to reposition one of my audio recorders, hoping to capture any other anomalies the night might offer. While Bondsy, Sarah, and Tuna-melt stayed behind at the base camp, I made my way back toward the old oak tree that stood like a sentinel in the heart of the graveyard. Twenty yards beyond that tree lay a peculiar clearing, an area devoid of headstones. For the longest time, this spot puzzled me. It seemed like the ideal place for a burial, yet it stood empty, untouched by markers of the dead.

My curiosity about the clearing nagged at me for years until a conversation with a local resident—a man who had spent his entire life on Williamsburg Hill—shed some light. He told me that the clearing was no ordinary piece of land; it was sacred ground. Native Americans had been buried there, their graves once marked by stones that had since been swallowed by the earth, hidden by decades of soil erosion. The revelation was chilling, yet it made perfect sense in retrospect.

Even before learning about the Native American burials, I had sensed that something was different about this part of the cemetery. There was an undeniable strangeness to it. At times, the air was cooler here, significantly cooler than in the rest of the graveyard, which defied logic since there were no trees to provide shade. The sensation was unnerving, as if the very ground beneath me was steeped in an ancient energy, a coldness that seeped into my bones.

On more than one occasion, while near these hidden graves, I had felt an inexplicable presence—an unsettling awareness of someone or something watching me. Other times, I felt as though I was being followed, but whenever I turned around, there was nothing but empty darkness. It's moments like these that remind me how thin the veil between our world and the next can be, especially in a place like this, where the past refuses to stay buried.

After crossing the cemetery, I carefully placed an audio recorder near a weathered grave marker just a few feet from where the Native American graves are said to be. It was a routine step, one I'd taken countless times before in other investigations. The moment itself felt still, uneventful, with nothing strange catching my attention. However, a few days later, as I sat down to review the audio, I was in for a surprise. What I discovered on that recording was beyond anything I had anticipated.

Among the ambient sounds of the cemetery, a clear EVP emerged—one of the most striking and undeniable pieces of evidence I've ever captured. It wasn't the usual faint, ambiguous whisper or murmur that could be easily dismissed. No, this was something different. It was a chant, rhythmic and unmistakable, resonating with a haunting clarity. Phonetically, it sounded like "Ye-ya-he-he," as though an unseen presence was performing an ancient ritual. The chant was accompanied by a single, human-like whistle, both before and after the eerie melody

What made this evidence so compelling wasn't just the clarity of the chant, but its context. We hadn't heard it during the investigation—not a single one of us. But here it was, captured on the recorder near the burial sites of those who walked these lands long before any of us. The connection was too strong to ignore, too specific to dismiss as mere coincidence. This wasn't just another EVP; it was a piece of history echoing through time, a direct link to the spirits of Native Americans who still, perhaps, keep watch over Williamsburg Hill.

It's moments like these that draw me back into the world of the unexplained and remind me why I started down this path in the first place. The paranormal has a way of offering glimpses into realms we can barely understand, and each encounter leaves me with more questions than answers. But with evidence like this, it's hard to dispute that something extraordinary is at play. And so, the quest continues, always in search of the next clue, the next whisper from the beyond.

Earlier in the chapter, I mentioned wanting to try an experiment at the boy's grave with the girls. When Sarah walked in front of the grave, I asked her to stop and stand there for a moment, curious to see if she would feel anything. After a few seconds, she shrugged and said, "I don't feel anything."

"You don't feel anything?" I pressed, almost surprised.

"No, nothing at all," she repeated, her tone even.

Since Sarah didn't react the way I thought she might, I asked her to step aside and called Taylor over. The moment she stood beside the grave, without any prompting, she blurted out, "I don't know what it is, but I'm doing all I can to keep from crying!" Her voice trembled, and before she moved away, she had to wipe tears from her eyes.

Bondsy, watching all of this unfold, looked at me with a mix of disbelief and suspicion. "No way! You told her!" he exclaimed.

"I didn't tell her a thing," I replied calmly. "I didn't even meet Taylor until tonight, and I certainly didn't say anything to Sarah."

As the investigation continued, I found myself pondering the strange effects the boy's grave had on Taylor but not on Sarah. A few days after our trip to Williamsburg Hill, Bondsy and I had a conversation that shed some light on this mystery. He offered a theory that made a lot of sense to me. He believed that the stories I shared during our drive—tales of run-ins with drug dealers in the parking lot and legends of panthers roaming the hill—had left Sarah so anxious about the real-world dangers lurking in the woods that she was too distracted to notice anything paranormal. In other words, her fear of physical threats may have overpowered or blocked any potential emotional reaction to the grave.

Taylor, on the other hand, was initially just as uneasy about being in a haunted cemetery, perhaps even more so. But something changed when she stood by that boy's grave and was overcome with emotion. Later, Taylor confided in me that after the intense feelings she experienced at the grave, a sense of calm washed over her, dispelling her earlier fears. As she described this sudden sense of peace, I couldn't help but wonder if the voice we'd heard through the spirit box earlier—the one that seemed to be her grandfather—had something to do with it. Perhaps it was his presence, rather than the boy's grave, that had calmed her nerves.

Whatever the cause, there was no denying that something extraordinary had happened that night, something that left a lasting impression on all of us. Taylor's experience, in particular, was a powerful reminder that the paranormal is as much about the emotions it stirs within us as it is about the strange phenomena we encounter. And in this case, those emotions seemed to hold the key to understanding the true nature of the events at Williamsburg Hill.

As Bondsy reviewed more of the audio from that night, he stumbled upon something unexpected—a second EVP captured near the boy's grave. This time, it was on the recorder he had brought from the station. At first, he almost dismissed it, thinking it was some kind of electronic interference that had distorted one of our voices. But what he found was far more unsettling.

The voice on the recording sounded strangely mechanical, reminiscent of the iconic "Darth Vader" voice from *Star Wars*. It was deep, distorted, and unlike anything else we had heard that night. Bondsy was puzzled, so he decided to play it for one of the station's technicians—a guy who had logged countless hours recording and editing audio and was familiar with the equipment we had used.

The technician listened carefully, noting that while our voices on the recording were clear both before and after the mysterious one, the unaccounted-for voice was inexplicably distorted. He couldn't wrap his head around it. If it were some kind of interference affecting the recorder, all of our voices should have been affected— but they weren't. Only this strange, mechanical voice stood out, further deepening the mystery of what we had encountered that night.

Bondsy had emailed me the audio file of the EVP, but I struggled to make out what was being said—none of us could. The voice was so distorted and mechanical that it was nearly impossible to decipher. When I arrived at the station on the morning of the Halloween show, I was unaware that Bondsy had taken the initiative to edit the clip using professional-grade audio editing software. He had slowed it down and removed as much background noise as possible, hoping to bring the voice into clearer focus.

As we approached the moment to air the clip, Bondsy explained that since our last conversation, he had worked on the file and believed that the adjustments he made would make the message understandable—even to me. My anticipation grew as he prepared to play it live.

However, as the clip played, my initial excitement quickly turned to disappointment. Despite the enhancements, I still couldn't make out the words. During the next commercial break, I pulled Bondsy aside and told him that the clip still seemed too fast and that it might need to be slowed down even more.

Bondsy went back to the editing software, made further

adjustments, and played the clip for me again. This time, even though the voice still had that strange mechanical quality, it was just slow enough that I felt like I could finally grasp what it was saying—or at least, I thought I could.

What I heard through the edited audio clip sent a chill down my spine. The distorted, mechanical voice seemed to be saying, "It's Larry's girl." The words were eerie, as if the voice were answering a question posed by another entity. Despite the unnatural tone, there was a distinct clarity in the message, and it unmistakably sounded like it was saying, "It's Larry's girl!" The significance of this phrase hit me immediately, but its implications were even more unsettling. Was this a reference to someone we knew, or was it connected to the grave we were investigating? The possibilities were both fascinating and deeply unnerving.

The phrase "It's Larry's girl" struck a deep chord with me, sending a wave of emotion that I hadn't anticipated. In 1997, my wife and I suffered an unimaginable loss—our baby girl, Kerri. The words echoed in my mind, aligning with something that had stayed with me since 2008, when I visited a psychic medium for the first time. She didn't know me, had no prior knowledge of my life, yet she seemed to see straight through to my soul. Among the many things she revealed, one stood out above the rest: she told me I had a daughter, an angel baby, who was with me constantly, watching over me.

I remember being stunned by her accuracy, particularly since she knew nothing of my past yet spoke as if she had lived it with me.

Her words had a profound impact on me, and I took them to heart. If what she said was true, and I truly had an angel baby by my side, then it wasn't a stretch to believe that other spirits could see her too. The statement from the EVP, "It's Larry's girl," suddenly made sense in a way that was both comforting and eerie. Could this be validation from the other side, a recognition of the presence of my lost daughter? The idea was both heart-wrenching and oddly reassuring, adding another layer to the mystery of that night and the inexplicable events that seem to follow me.

Now we fast forward to 2020, the year I published my book *Strange Williamsburg Hill*, which delves into my extensive investigations of the hill and its enigmatic graveyard. This book explores the myriad of strange phenomena that have unfolded in that unsettling location. Among the numerous accounts, I dedicated significant space to the young boy's grave and the peculiar reactions it elicits from women. I recounted several incidents that I had personally witnessed at his gravesite, weaving them into the broader narrative of the cemetery's eerie reputation.

After the book's release, I received an intriguing message from the cousin of the young boy, who had read the book and shared it with the boy's mother. She informed me that many of the details I described resonated deeply with her, as they matched up with personal anecdotes about her cousin. At one point, she referred to him as "Thaddy," a nickname that, until then, I had not been aware of.

This revelation prompted me to revisit the collection of EVPs I had amassed over the years from Ridge Cemetery. When I played the clip that I had previously thought said, "It's Larry's girl," I was stunned to hear the clear phrase, "It's Thaddy's girl." The coincidence was jarring—it was as though the voice on the recording had been directly responding to the very details the cousin had confirmed, bridging the gap between the world of the living and the spectral realm in a way that was both profound and perplexing.

The EVP clearly connects to the young boy, but the lingering question is whether the voice belongs to him or if it's another spirit referencing him. I may never uncover the definitive answer to this puzzle, but the recording stands as compelling evidence of the extraordinary, if not supernatural, occurrences surrounding the boy's grave. For me, it underscores the reality that something truly unusual is at play in that hauntingly mysterious corner of Ridge Cemetery.

Investigation Summary

The investigation at Williamsburg Hill yielded a series of extraordinary experiences, with a surprising focus on Taylor. But why was she so uniquely affected in a place she had never visited before? This question adds another layer to the enigma of Williamsburg Hill, and much like many aspects of the paranormal, it may remain unanswered.

The day following our investigation, Taylor sent me a private Facebook message that revealed her confusion and lingering bewilderment. Considering she had no prior experience with paranormal investigations and likely expected nothing out of the ordinary, her sudden encounter with phenomena straight out of a Stephen King novel was understandably disorienting.

Picture this: in the dead of night, in a remote graveyard, your nickname is called out through a radio device. Even I, a seasoned skeptic of the SB-7 box, found it hard to dismiss the profound impact of such an experience. Witnessing the paranormal firsthand can be a transformative event, and it's something Taylor, Bondsy, and Sarah would likely attest to with conviction.

If Taylor's nickname had been something common, like Mary or Sue, one might easily attribute it to radio skip or mere coincidence. But "Tuna" is a name that stands out. To hear it come through the device, recorded not once but nine times, challenges any attempt to rationalize it away. It's moments like these that compel one to question the boundaries of our understanding and the reality of the unseen forces that lurk in the shadows.

In Taylor's message to me, she revealed her internal struggle about whether to share her unsettling experiences with her mother, Lindsey. She hesitated to disclose how a simple radio device, the SB-7 spirit box, had not only called out her nickname multiple times but also answered a deeply personal question about the color of her hair as a child. Taylor was even more reluctant to suggest that the voice might have belonged to her grandfather, Dandy, her mother's father.

Eventually, Taylor shared her bizarre experiences with Lindsey. She recounted the emotional turmoil she felt while standing by the boy's grave and described the spirit box's accurate response to her childhood hair color question. Taylor asked her mother if she believed that Dandy would be the type to reach out from beyond the grave, using any means available to communicate with the living.

Lindsey's response was both surprising and reassuring. She revealed that she had recently visited a clairvoyant, who had informed her that her father, Dandy, was always present, watching over them. "So yes, Taylor," her mother affirmed, "I absolutely believe that Grandpa Dandy would try to contact his granddaughter through a device like that if he had the chance."

The conversation took an even more eerie turn when Lindsey added, "I'm not sure if you realize this, but last night, when you were in the cemetery, was Grandpa Dandy's birthday!"

Reading this revelation sent an icy shiver down my spine. The convergence of these elements—the timing of the investigation and Dandy's birthday—was too coincidental to ignore. The most perplexing aspect of the voice on the spirit box remains: why did it address Taylor by her nickname rather than her given name? This question is a mystery that adds to the already profound and enigmatic experience, one that may never be fully explained.

I can't say for sure whether Bondsy and the radio station crew found the definitive proof they were seeking to confirm the existence of the paranormal. But what I do know, without a shadow of a doubt, is that the events of September 28, 2012, left an indelible mark on all of us—a night none of us will soon forget.

In Shelby County, the tales surrounding Williamsburg Hill's Ridge Cemetery have taken on a life of their own, feeding the legend that surrounds this eerie patch of land. For some, it's just another old graveyard, a place of quiet rest. But for others, it's something far more unsettling—a place that embodies fear, a location where the boundary between the living and the dead seems unnervingly thin. For me, Ridge Cemetery is a site of inexplicable occurrences, a mysterious nexus of paranormal phenomena that I'm drawn to, a place I hope to continue exploring for years to come.

Canadian author Robertson Davies once remarked, "The eyes see only what the mind is prepared to comprehend." While his words ring true in many contexts, I'd wager that Mr. Davies never had the occasion to visit Williamsburg Hill's Ridge Cemetery—a place where comprehension is continually challenged, where the line between what we think we know and what we truly understand blurs in the shadows. Ridge Cemetery is a veritable stew of paranormal activity, simmering with unexplained phenomena that many have borne witness to, yet few can fully grasp.

As for me, I can only hope that I'll be fortunate enough to continue delving into the mysteries of this place, to witness more of its strange offerings, and perhaps, one day, to unravel the enigma that cloaks its haunted grounds.

But for those who feel the urge to visit, to confront the unknown, a word of caution: the monsters that inhabit the darkest recesses of our minds often find ways to manifest in the real world. So, when you arrive at Ridge Cemetery, take heed—park your car facing the exit. You never know when you might need to make a quick escape.

GRANITE CITY YMCA
GRANITE CITY, ILLINOIS

TWO

Who's watching
Tell me, who's watching
Who's watching me
- Pop Artist, Rockwell

There are moments when a certain song on the radio takes me back to the field, conjuring vivid memories of past investigations. One track in particular, Rockwell's 1984 hit "Somebody's Watching Me," has an uncanny way of transporting me back to a specific night in July 2013. It was an investigation at the old Granite City YMCA in Granite City, Illinois. The lyrics, dripping with paranoia and unease, perfectly capture the atmosphere of that evening. But what we encountered there went beyond the eerie sensation of being watched. We weren't just under observation; something—some unseen force—was actively engaging with us. It made its presence known in the most unsettling way possible—by hurling objects in our direction—a stark reminder that in places like these, the past is often far from silent.

Granite City, nestled in Madison County, Illinois, lies just seven miles east of the bustling St. Louis metropolitan area. Officially founded in 1896, its history stretches back far beyond that date. The land where the city now stands was originally settled by the French, followed by waves of German and English settlers, long before it became the industrial hub it's known as today. The city's fortunes were closely tied to its location along the Mississippi River, a vital artery during the Industrial Revolution. It was this proximity that attracted local entrepreneurs Frederick and William Niedringhaus, who established the St. Louis Stamping Company—a venture that would eventually evolve into the massive Granite City Steel.

In the early 1800s, settlers were drawn to the fertile lands just east of St. Louis, and by the 1830s, the area had become known as Six Mile, named for its distance from the river city. By 1896, Granite City had officially incorporated as part of Madison County, and its industrial roots began to take firm hold.

The Niedringhaus brothers played a pivotal role in shaping the city's growth. Their success led to a rebranding of their company as the National Enameling and Stamping Company, a name change designed to appeal to a broader market. But it wasn't just their business acumen that brought prosperity; it was the promise of steady, well-paying jobs that drew waves of European immigrants from Macedonia, Hungary, and Bulgaria, all seeking a better life in America.

Many of these immigrants settled in a neighborhood west of downtown that would later become known as Hungry Hollow. This area of town was no stranger to hardship, with the first major tragedy being the flood of 1903. The floodwaters submerged much of what is now the western part of Granite City, and local lore tells of immigrants, living in makeshift tents, who were swept away by the deluge—some of whose bodies were never recovered. It's a haunting story, though one I must note remains unconfirmed and undocumented in the historical record. Yet the grim realities of life in Hungry Hollow were undeniable. As jobs dwindled, many of the immigrants faced starvation, with some succumbing to hunger in their search for the American dream.

On August 28, 1924, the first shovelful of earth was turned, marking the beginning of construction for the Granite City YMCA. By May 11 of the following year, the cornerstone was laid, and on January 3, 1926, the doors officially opened to the public. This grand structure, a beacon of community and fitness, soon became a hub of activity for the residents of Granite City. In 1954, the building was expanded with a new wing and viewing area, housing racquetball courts and providing more space for its growing membership. The YMCA even played host to a young Kevin Cronin and the band REO Speedwagon, who performed a concert in its gymnasium before they became rock legends.

The YMCA stands at 2001 Edison Avenue, nestled in a primarily commercial neighborhood. The building's location is no coincidence—it sits in an area of town steeped in history, where tragedy struck the immigrant communities that settled there in the early twentieth century. Unfortunately, the ground on which the YMCA was built seems to have absorbed some of that tragic energy, as the building itself has been the site of several alleged misfortunes.

Over the years, rumors of death and despair have circulated around the YMCA. One story that has persisted is that of a young girl, somewhere between the ages of nine and twelve, who drowned in the swimming pool while playing with other children. Another tragic tale involves the murder of a man on the front steps of the YMCA by a fourteen-year-old boy during a botched robbery attempt. There's also the haunting story of a former janitor who was mentally challenged and reportedly took his own life in a small work area near the old locker rooms. Heartbroken over unrequited love for a female member of the YMCA, the man allegedly hanged himself in despair.

The tragedies don't stop there. Two male members of the YMCA are said to have died from heart attacks on the premises—one in the men's locker room and another named Andrew, who was sixty-three years old when he passed away while watching a movie in the dayroom. There's even a vague report of a death in Room 207, though the details remain murky.

It's important to note that I have not been able to confirm or document any of these alleged deaths through official records. As with many urban legends, the truth can be elusive, shrouded in the passage of time.

The 1111 Phenomena Strikes Again.

For those familiar with my work or who have read my previous books, you already know about the peculiar pattern that has followed me since I began my journey into the paranormal—the recurring appearance of the number sequences 11 and 1111. These sequences don't just pop up occasionally; they bombard me with such frequency that it borders on the surreal. It's not just a random glance at a clock or a fleeting sight of a license plate bearing the number 1111. It's more pervasive than that, like the time I received a text message while driving in my car at exactly 11:11 a.m., only to realize I was reading a text message sent at 11:11 p.m. the night before. When I glanced up to check on traffic, I had to slam on my breaks as a car had stopped in front of me. I stopped in the nick of time before rear-ending the car. As the car pulled away, I almost fell over because the license plate number of the car was, 1111. The odds of such things happening in such a precise, synchronized manner must be astronomical.

Not long after I started noticing these repeating sequences, I also began to experience an uptick in synchronicities—those seemingly coincidental events that skeptics might brush off as mere happenstance. But if you know me, you know that after years of delving into the paranormal, I've come to believe that coincidence is just a term we use when we can't yet see the connection.

This brings me to a story that underscores the significance of the 1111 phenomenon.

As an investigator, I'm always on the lookout for new locations to explore, particularly in the Midwest, where history is rich and the paranormal thrives. The way I gained permission to investigate the old Granite City YMCA, however, was nothing short of uncanny. I started by conducting a Google search using the keywords "haunted" and "Illinois," hoping to uncover a new and intriguing location.

Among the results, I stumbled upon a webpage belonging to a paranormal group based in Quincy, Illinois. The page listed various locations they had investigated, including the YMCA in Granite City.

As fate would have it, I recognized several members of this Quincy group from a presentation and book signing I had done for them in the past. It seemed like a natural next step to reach out to one of them for contact information or a lead on how to gain access to the Granite City YMCA. However, I was knee-deep in reviewing evidence from a recent investigation, so I decided to put off contacting the Quincy folks for the time being.

Meanwhile, my curiosity about the YMCA persisted, so I continued researching the building. During this deep dive, I came across several newspaper articles detailing investigations conducted by a local paranormal group from Granite City. One article stood out; it mentioned a man named Bill, who was listed as the point of contact for anyone interested in investigating the old YMCA. The article even provided his phone number.

And here's where the 1111 phenomenon made its presence known once again. Bill's phone number ended in 1111. When I saw those digits staring back at me, I knew this wasn't just a coincidence. It felt like a nudge from the universe, a sign that I was meant to explore whatever mysteries the Granite City YMCA held within its walls.

A Call from Beyond Coincidence

Several weeks had passed, and despite my initial resolve, I still hadn't made the call to Bill. I distinctly recall that it was a Tuesday morning when I finally decided to reach out to him during my lunch break to see if I could schedule an investigation at the YMCA. But as fate would have it, a friend invited me to lunch that day, leaving me with no time to make the call. I promised myself I'd do it on the drive home.

As I settled into my car later that afternoon, I picked up my cell phone, ready to dial Bill's number and introduce myself. That's when I noticed I had a missed call. Curious, I selected the option to display the number, and what I saw made my heart skip a beat: "###-###-1111."

I immediately recognized the number because of the distinct area code and, of course, the 1111. But how? And more importantly, why would Bill be calling me? He didn't know me, had never met me, and, as far as I was aware, didn't have my cell phone number.

Without hesitation, I hit the redial button. Bill answered, and after a brief exchange of pleasantries, I couldn't help but share with him the strange coincidence—or what some might call synchronicity—of the situation. I told him how I had been planning to call him for weeks to arrange an investigation at the YMCA, and on the very day I finally committed to making that call, he ended up reaching out to me first.

Bill explained that he had recently been in touch with the Quincy paranormal group, the same team that had investigated the YMCA before. During their conversation, they mentioned my name and provided him with my phone number. It turned out that Bill was in the midst of organizing a paranormal conference to raise funds for the restoration of the old YMCA, and when he told the Quincy group he was looking for guest speakers, they recommended me.

As we continued to talk, Bill asked if I would be interested in speaking at the fundraising event. I agreed, and when he inquired about my speaking fees, I told him that I don't charge for fundraisers. Grateful for the offer, Bill then said that in return, he would allow me access to the building to conduct my investigation free of charge. We wrapped up the conversation with an agreement that I would call him back in a few days to confirm the date for the investigation.

And so, a few days later, I reached out again, and we settled on the date—Thursday, July 25, 2013.

It was one of those moments when you realize that some things are just meant to be, where the line between coincidence and the inexplicable blurs into something that feels eerily predestined.

When I told Bondsy about the old YMCA and explained the history of the building, he was all for an investigation there. Since Granite City was only ninety minutes away from the station, we figured it would be the perfect place.

Our plan was to kick off the investigation at 10:00 p.m. on the night of July 25, wrapping up around 4:00 a.m. the following morning. Since this investigation involved the radio station, the team was larger than usual. Alongside Bondsy, his co-host Sarah Savannah, and two female interns from the station—nicknamed Trigger and Lil Beans—would be joining us. Neither of the interns had ever been on a paranormal investigation before, which always adds an unpredictable element to the night. We were also joined by Chris, a seasoned paranormal investigator I had mentioned earlier in the book.

The plan was straightforward: I would meet Bondsy and his radio team in Springfield at 8:00 p.m., and from there, we'd make the nearly two-hour drive to the YMCA building in Granite City, where Chris would be waiting for us.

As I drove to Springfield to pick up the radio crew, I decided to call Bill to confirm our arrival time at the YMCA. But just as I reached for my phone to dial his number, the phone rang. The timing was so perfect it was eerie. When I glanced at the screen, a familiar feeling of unease washed over me as I recognized the number: ###-###-1111.

It was the same uncanny sequence that had caught my attention before. Seeing it light up my phone again made the hair on the back of my neck stand up. In that moment, I knew this investigation was going to be something extraordinary.

Bill was calling for the same reason I had intended to call him— to confirm our arrival time. The synchronicity of it all was hard to ignore, as if the investigation was already steeped in the strange energy we were about to encounter.

After picking up the KISS FM crew at the station, we headed south on Interstate 55 toward Granite City, arriving just before 10:00 p.m. Chris had beaten us there by a few minutes, waiting in his car as we pulled up.

We decided to leave our equipment in the vehicles for the time being and entered the building, where we were greeted by Bill and several members of his team. After brief introductions, Bill and one of his colleagues took us on a tour of the building. As we walked through the shadowy halls and dimly lit rooms, I began to feel a familiar, unsettling sensation—a strange tingling at the base of my spine that I'd learned not to ignore.

The first location that triggered this feeling was the old locker room and pool area, a space that seemed to hold its breath in the darkness. The second was on the second floor, where the air felt thick, as though it was steeped in memories of the past. I couldn't shake the feeling that something was waiting for us in these places, something that would make its presence known before the night was through.

Bill led us to an area near the racquetball courts, where piles of old drywall littered the floor, remnants of ongoing renovations. He explained that they were working to clean up the building, making it a safer and more pleasant environment for visiting investigation groups. But then, with a serious tone, he pointed his flashlight at the piles of debris and warned us to be cautious where we walked and sat. The reason for his caution became clear as the beam of light revealed hundreds, if not thousands, of Brown Recluse spiders crawling among the rubble.

These spiders are not to be taken lightly—they're venomous, and a bite can cause serious health issues, even necrosis of the skin. The sight of them swarming over the debris added an extra layer of tension to the night, as if the building itself was laced with hidden dangers, both seen and unseen.

As we continued the tour, the weight of the night ahead settled on us. The eerie feeling in the locker room and on the second floor lingered in my mind. We hadn't even unpacked our gear yet, but already the building was beginning to reveal its secrets. I had the distinct sense that the night was going to be one for the books, filled with the kind of activity that would keep us on edge and test our resolve as investigators.

After witnessing the unsettling sight of the spider infestation, Bill led us deeper into the bowels of the building, guiding us to the old locker room and pool area. The air here was heavy with the kind of silence that seemed to listen, as if the walls themselves were waiting to reveal their secrets. Bill paused near the pool, his voice low as he recounted a tragic story that had become part of the building's dark lore.

Years ago, a little girl, playing with friends, had somehow drowned in this very pool. Since then, visitors to the building have reported hearing the disembodied voice of a child echoing through the pool area—a soft, fleeting sound that leaves a chill in its wake. Just behind the pool was a small workroom where, according to local legend, a former janitor had taken his own life, hanging himself in despair. The stories were enough to make the hair on the back of your neck stand up, even before you heard the first ghostly whisper.

From there, we moved to the second floor, where the atmosphere grew even more oppressive. Bill explained that many visitors feel a deep sense of unease on this floor, as if they're being watched or followed by something unseen. The rumor that someone had died in Room 207 lingered like a shadow over this part of the building, though no records existed to confirm the story. But as the night would soon prove, documentation wasn't necessary to feel the weight of whatever lingered here.

During the walkthrough, several members of our team began to feel the same unease. Bondsy, in particular, seemed affected, his usual bravado tempered by a growing discomfort. The rooms near the staircase, with their dim lighting and oppressive silence, seemed to press in on him, amplifying his anxiety. Little did he know, this was only the beginning of the uneasiness he would face tonight.

As we finished the tour, I couldn't shake the feeling that the building was aware of our presence, as if it was waiting to see what we would uncover. The stories of tragedy and the discernable tension in the air set the stage for what promised to be an unforgettable investigation, one that would test not just our courage, but our very understanding of the paranormal.

After Bill concluded the tour, we made our way back to the first-floor lobby, a cavernous space that would serve as our basecamp for the night. The lobby, with its high ceilings and echoing footsteps, seemed to hold its breath as we prepared for the investigation ahead. We wasted no time bringing in our equipment, the hum of anticipation hanging thick in the air.

Once everything was in place, Bill and his crew bid us goodnight, leaving us alone in the building. The door closed behind them with a heavy thud, and suddenly, the full weight of the empty YMCA settled around us. The silence was almost tangible, broken only by the occasional creak of the old building settling into the night. It was just us now, our gear, and whatever might be lurking in the shadows of this place, waiting to reveal itself. The real work was about to begin.

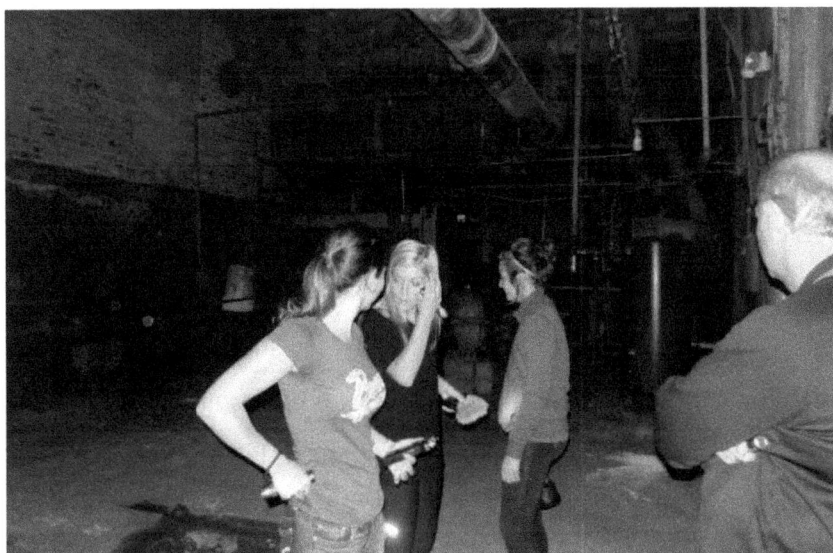

The Swimming Pool area

Given the uneasy vibes that had gripped our team, we decided to begin the investigation on the second floor, near the staircase where the strange feelings seemed to be most concentrated. The sense of discomfort was palpable, especially for Trigger and Lil Beans, our two interns, who were clearly on edge. Their anxiety was understandable—this was their first dive into the world of paranormal investigation, and the weight of the unknown was pressing down on them. They stuck close to the rest of us, hesitant to stray too far from the safety of the group.

As we began our exploration, the silence of the old building was unnerving, but nothing out of the ordinary seemed to be happening. The air was thick with anticipation, yet no signs of activity emerged. Sensing a lull, I decided to change tactics. Lil Beans and I broke off from the main group and headed down the hallway toward the racquetball courts, leaving Bondsy, Sarah, Trigger, and Chris to continue their investigation near the staircase.

Upon entering the racquetball area, Lil Beans suddenly stiffened. She claimed to hear faint voices—murmurs just on the edge of perception. She insisted they were different from the voices of our teammates, almost as if they were whispers from another time or place. I strained to hear them myself, but the silence in my ears was deafening. Lil Beans continued to pick up the strange murmurs for several minutes before they abruptly stopped, leaving us in an eerie stillness.

We decided to step back into the hallway, hoping to catch the voices again or at least get a better sense of what we were dealing with. But as we paused in the hallway, our thoughts were interrupted by the sudden sound of raised voices from our fellow team members. Their excitement was unmistakable, echoing down the corridor. Whatever they had encountered, it was significant enough to jolt them out of their quiet investigation. Without a second thought, we hurried back down the hall, eager to find out what had stirred them up.

When we caught up with the group, the tension was thick in the air. Bondsy turned to me, his eyes wide with a mix of disbelief and excitement.

"Dude, you're not going to believe what just happened!" he blurted out.

The urgency in his voice made it clear that something out of the ordinary had taken place. Sarah stood beside him, noticeably calmer than the others, but then again, she had accompanied me on investigations before and was no stranger to the strange and unexplained.

Bondsy quickly recounted the incident. While Chris and Trigger were investigating a room at the end of the hall, he and Sarah were standing near the staircase when they heard a sudden, startling noise. According to Bondsy, it sounded as if something had been thrown in their direction. They immediately started searching for the object but had come up empty-handed. He described the sound as a series of hollow metallic pings—like something bouncing off the floor multiple times.

"It was like a ping, ping, ping," he said, his voice still tinged with apprehension.

The expression on Bondsy's face told me everything I needed to know—this had shaken him. Even before the mysterious object had made its appearance, he had been on edge, a state of mind entirely out of character for him. I had never seen him this rattled before, and it was clear that whatever had just transpired was more than enough to push him to the brink of his comfort zone.

We continued our search, determined to identify the object that had caused the unsettling noise. After several minutes, Chris discovered an empty .22-caliber shell casing on the floor. Hoping this might be the culprit, he tossed it down the hallway. The sound it made, however, immediately struck the group as different. It lacked the distinct, hollow ping that had echoed through the corridor earlier. Even after kicking the casing around—just in case one of us had accidentally sent something flying with a stray foot—the conclusion was unanimous: this wasn't the sound they'd heard.

For fifteen minutes, we scoured the area, but the mysterious object that had startled Bondsy and the others remained elusive. Frustrated but intrigued, we decided to take a break and headed back down to the first floor. Our next move was to review the footage on the video camera to see if it had captured any clues about what had just happened.

Once downstairs, Sarah handed me the camera. I rewound the video to just before the team had heard the noise, my hopes high that Sarah's steady hand—she had been walking slowly to minimize camera shake—had provided us with clear evidence of the incident. And indeed, the video did not disappoint.

The footage showed Trigger standing beside Bondsy, taking temperature readings with a thermometer. She noted that the room temperature was fluctuating between 80 and 81 degrees. Bondsy mentioned how the temperature had been 89 degrees only minutes earlier, indicating a significant drop of eight degrees. Curious to document this, Bondsy asked Sarah to bring the camera closer to capture a shot of the thermometer's LCD screen.

As Sarah was recording the temperature, the noise rang out from the end of the hallway. Both Chris and Bondsy reacted instantly, their voices overlapping as they exclaimed, "Did you hear that?" Without hesitation, Chris began to move down the hallway, intent on locating the source of the sound.

Once he reached the end of the hall, Chris, directing his words toward whatever unseen presence might be with us, called out, "Can you knock on the door of the room you're in, or tap on something, so I know where you are?"

Bondsy, still processing the situation, asked Trigger to join Chris at the end of the hallway to check for any additional temperature fluctuations. But whether it was due to fear or some other reason, she hesitated and ultimately did not follow his request. The tension was unmistakable, and the mystery of the unexplained noise only deepened the unease that had already settled over the team.

As I reviewed the footage, one thing struck me: Bondsy didn't follow Chris down the hallway after the noise was heard. This was unlike him. When I asked him why, he admitted that something about the second floor unsettled him. The sound had unnerved him so much that he simply couldn't bring himself to walk to the end of the hallway.

The video continued, showing Chris entering the last room on the right for a few seconds before returning to the hallway. As he emerged, Bondsy, still clearly uneasy, asked Trigger once more to join Chris and check the temperature reading. This time, she complied, though her hesitation was apparent. Bondsy instructed

Sarah to stay where she was, a subtle hint in his tone revealing just how uneasy he felt. He clearly wanted someone nearby, someone to anchor him in the growing tension.

Then something unexpected happened. As Chris turned toward the group and took a step forward, the camera captured a faint flash of light at the bottom of the wall behind him. The flash was quick, almost imperceptible, but it was there. Chris reacted immediately, turning around and asking, "Did you guys take a picture?"

Trigger's response was immediate: "No!"

Chris pressed on. "Did you see a flash behind me?"

"I did," Sarah replied, her voice steady. "I thought you took a picture."

Chris's voice, now tinged with excitement, confirmed the strangeness of the moment. "No, I didn't."

The exchange left the team questioning what they had just witnessed. A simple flash of light, but in the context of where we were and what we were doing, it was anything but simple. It was another layer of the unexplained, adding to the growing sense that something was with us, watching, interacting, and perhaps even warning us as we continued our investigation.

As Chris felt a sudden and bone-chilling drop in temperature, he called the team to join him at the end of the hall. Trigger was the first to respond, quickly moving to Chris's side with the infrared thermometer in hand. Chris's discomfort was evident—he pointed to the goosebumps on his arm, his voice trembling as he explained the sensation of intense cold.

Trigger checked the thermometer, and her voice, visibly shaken, reported, "It's 22 degrees." The room had plummeted 67 degrees from the temperature recorded just minutes earlier, an astonishing drop that underscored the eerie nature of the evening.

The phenomenon of extreme temperature drops during paranormal investigations is well documented and widely reported among those who delve into the unknown. Many believe that ghosts or spirits draw on the ambient heat to manifest their presence, effectively chilling the environment as they materialize. This theory assumes that the heat energy from the surroundings is siphoned off by the entities, resulting in a sudden and noticeable drop in temperature.

Having spent over two decades investigating the paranormal, I have my own perspective on these chilling occurrences. I contend that extreme temperature drops might be linked to the opening of portals or interdimensional windows. When entities from other realms cross into our reality, they momentarily bring with them the temperature from their own dimension. Alternatively, these spectral beings might be so devoid of heat that their presence alone causes a measurable cooling effect in our environment.

As the team gathered around, the temperature drop was more than just a curious anomaly—it was a clear sign that something strange was unfolding in the old YMCA.

As Sarah began her cautious approach down the hallway toward Chris, the video footage captured every deliberate step. Bondsy, trailing behind her, voiced his growing discomfort. "Don't go too far," he urged, a note of unease evident in his voice. When Sarah inquired about the cause of his concern, Bondsy bluntly admitted, "Because I'm getting creeped out!"

This investigation at Granite City was only Bondsy's second paranormal foray with me. Since then, he has participated in numerous investigations, but none have rattled him quite like the unsettling experience at the YMCA's second floor. Bondsy's reaction was out of character, a stark contrast to his usual demeanor during investigations.

Despite Bondsy's trepidation, Sarah was determined to document any potential paranormal activity. "If something shuts a door, I want to record it," she asserted, resolutely continuing her advance down the corridor.

The video footage from this moment is particularly revealing. As Sarah took her second step, the camera's gentle up-and-down motion captured a distinct sound—a metallic, coin-like clatter that reverberated down the hallway. It was the unmistakable noise of something bouncing on the concrete floor, skimming and ricocheting three times.

Chris's voice cut through the tension. "Did you do that?" he asked, directing his question at Sarah. Bondsy chimed in, his voice tinged with a mix of concern and accusation, "Sarah, you kicked something," to which Sarah quickly responded, "No I didn't!"

The ensuing footage captured a priceless moment of raw emotion. As Sarah's denial echoed in the hallway, Trigger's reaction was swift and telling. She turned on her heel and, with a mixture of a half-grin and visible fear, hurried toward Sarah. Her expression conveyed a clear message: she was done with this unsettling experience.

Bondsy's unease was evident. Every few minutes, he would mutter about how "creeped out" he felt, the seriousness in his voice impossible to ignore. I could see the strain in his eyes; it was as though the shadows on the second floor had somehow seeped into his soul.

Chris, meanwhile, had turned his flashlight beam toward the floor, methodically scanning for the source of the strange noise. His search yielded an empty .22 caliber shell casing, which he tossed down the hall in an attempt to replicate the sound. The resulting clatter was a poor echo of the original, bouncing noise. The shell casing made a dull thud, lacking the metallic ping-pong resonance that had unsettled the group earlier.

Frustration was beginning to set in, so we decided to take a break. Sarah, Trigger, Lil Beans, and Chris headed down the street to a local convenience store to grab water and soft drinks, leaving Bondsy and me behind to monitor the equipment. Bondsy's tension was almost tangible, his reluctance to revisit the second floor evident in every word he spoke.

With the team out of sight, I decided to make my way back to the second floor. Bondsy, still visibly shaken, was adamant about not joining me. "I'm not going back up there," he declared, his voice betraying a mix of fear and resignation.

I understood his apprehension well; I had faced my own share of unnerving experiences in the field. But, as any seasoned investigator will tell you, confronting the unknown is a fundamental part of the journey. I took a moment to address Bondsy's fears, trying to offer him some reassurance.

"Look, I get it," I said, trying to sound as reassuring as possible. "Some places just give off a vibe, a feeling you can't shake. But part of what we do is pushing through those feelings. If you avoid confronting what makes you uneasy, it might hold you back in your pursuit of the paranormal. Facing your fears, even when they're this intense, is a crucial step. Believe me, it'll help you grow stronger as an investigator."

My words seemed to resonate with him. Bondsy looked at me, the conflict in his eyes slowly dissolving into determination. After a few moments of hesitation, he nodded. "All right," he said, his voice firmer now. "Let's go."

We retraced our steps along the second-floor hallway, methodically searching for evidence of the mysterious object that had made its presence known with such startling clarity. We threw, bounced, and kicked every small item we encountered—a screw, a small rock, chunks of plaster, and a nail—testing each for its resemblance to the peculiar sound captured on our video. Nothing produced the right pitch or resonance until I came across a piece of glass. It was roughly two and a half inches long and half an inch wide.

Intrigued, I gave it a tentative kick. The sound was close, though not quite an exact match. Determined to find the source, I picked up the shard and tossed it a few feet ahead. The glass bounced several times, producing a sound nearly identical to what we'd heard in the video. The moment Bondsy heard it, his face lit up with recognition. "That's it," he exclaimed. "That's what we heard!"

To ensure we had ruled out all possibilities, I let the glass fall from my hand to see if it might have been lodged in the ceiling and dropped. While it produced a somewhat similar sound, it lacked the precise quality of the noise made when I had tossed it. The consistency suggested that the object making the sound was indeed a piece of glass.

But the mystery remained: who or what had thrown it?

Shortly afterward, the team returned with refreshments, and we regrouped to share our findings. Bondsy and I briefed the others on the glass shard and our process of elimination. Then, while the team took a break, I headed back to the second floor to set up an infrared video camera on a tripod. The plan was to capture any further anomalies on film.

As I made my way back to the first floor, halfway down the stairs, a chilling whisper floated through the air: "Larry." The voice was clear, yet it carried an eerie, almost ethereal quality. I turned and looked toward the team at the bottom of the stairs, asking if anyone had called my name. They all responded with puzzled shakes of their heads. "No," they said in unison.

Unfortunately, no audio recorders were nearby to capture the whisper. The incident remained a ghostly fragment of our investigation, leaving me with an unsettling sense of having brushed against something unseen.

After our brief break, we reconvened and headed back to the locker room and pool area to resume our investigation. The building, cloaked in its late-night silence, seemed to hold its breath as we began our exploration. We meandered through the various rooms, the echo of our footsteps mingling with the creaks of the old structure. We eventually found ourselves at the edge of the swimming pool, a large, empty basin that seemed to stretch into the darkness.

Chris and I stood side by side near the pool, deep in conversation about the strategy for splitting up into three pairs. The rest of the team was positioned in front of us and to our right, their figures obscured in the dim light. As we debated the logistics of our next moves, an unsettling sound pierced the air—a distinct, shushing noise that seemed to come from the doorway behind us or from the room just beyond.

Instinctively, Chris and I turned toward each other and, almost in unison, asked if anyone else had heard that peculiar sound. To our surprise, we were alone in our experience, though this would soon change. Within the next hour, other members of the team would also report hearing the same eerie shushing.

Each occurrence of the shushing coincided with moments when the team was engaged in conversation, ruling out the possibility that the sound was made by anyone in our group. A few days later, when I reviewed the audio recordings from the swimming pool area, I discovered that the mysterious shushing had been captured not once but twice. This confirmed that not only had we heard it, but it had been documented as well.

Following the shushing sounds, we decided to split up for a more thorough investigation of the pool and locker room area. Bondsy, Trigger, Lil Beans, and Chris ventured into the shadows, their forms disappearing into the darkness as they began their search. Meanwhile, Sarah and I remained in the pool area, our focus sharpened by the eerie quiet that enveloped us.

I decided it was time to conduct an EVP session, aiming to contact the spirit of the little girl who, according to local lore, had tragically drowned in this very pool.

Sarah and I positioned ourselves at opposite ends of the pool, about fifteen feet apart. Sarah carried her tape recorder, while I placed mine carefully at the edge of the pool, facing inward. The obscure lighting cast eerie shadows across the empty pool adding to the atmosphere of our endeavor.

I began the session with my standard set of questions, hoping to elicit a response from any entities that might be present. I asked, "Who is here? Can you tell me your name? How old are you?" Each question was followed by a deliberate pause of about twenty seconds, providing ample time for any potential responses to be captured by our recorders. The stillness of the pool area was punctuated only by the distant voices of our team, and the occasional drip of water echoing in the empty space.

As we waited in the quiet, I remained attentive, keenly aware that even the faintest sound or shift in the atmosphere could be significant.

Sarah, Bondsy, and the Author by the pool

After several minutes of asking questions with no discernible response, I turned to Sarah and suggested, "If a little girl drowned here and her spirit is lingering, maybe she'd respond better to another girl. Why don't you take a turn asking questions?"

Sarah agreed and began her inquiry with a gentle, inviting tone. "Hey, little girl, do you like to play with Barbies?" As soon as she finished her question, her eyes widened in surprise. "I just felt something touch my leg!" she exclaimed, her voice tinged with both curiosity and alarm. I could barely contain my own astonishment as I replied, "No kidding." Then asked our invisible guest, "Do you like Sarah?" A few days later, while poring over the audio from my recorder, I was struck by what I found. Immediately following Sarah's question, a faint whisper emerged from the recording— barely audible, yet unmistakably clear—it seemed to say, "No."

Sarah's review of her own recorder yielded intriguing results as well. Among the usual ambient noises, she discerned a distinct voice, markedly different from the whisper captured on my device. It sounded unmistakably childlike, and as I listened carefully, I could make out the word "Mouse!"—a curious and seemingly unrelated reply, but one that added an eerie intrigue to our findings.

These snippets of audio suggested that, even in the absence of visual evidence, we had encountered something genuinely inexplicable during our investigation.

A few days after our investigation, I returned to the radio station, eager to share the results with Bondsy and Sarah. We gathered around the equipment as Bondsy prepared to play back the recordings. First, he cued Sarah's tape, and then mine, to compare the two sessions.

Bondsy, ever the technician, had a brilliant idea. He set up his audio editing software and created two separate tracks—one for Sarah's recording and one for mine. He meticulously aligned the two clips, so they played simultaneously. The result was striking. When played together, the whispers from both recordings formed a coherent "No!"

This discovery led to a tantalizing question: was this definitive "No" from the spirit of the little girl who reportedly drowned in the pool, or was it the voice of another phantom child wandering the building's eerie corridors?

After investigating the pool area, Sarah and I decided to head back to the second floor, where the glass had mysteriously been thrown earlier in the night. Bondsy and the rest of the team stayed behind in the locker room, a dimly lit maze of narrow corridors, doorways, and blind turns. It's a place where disorientation comes easy, especially in the pitch black. The team quickly realized they had left their flashlight near the pool, leaving them stranded in the darkness.

Chris had his digital camera with him, so Bondsy suggested using it to light their way back. "Can you take a flash burst with that camera?" Bondsy asked, his voice cutting through the thick silence. Chris agreed, and as he prepared to take a series of photos, a loud, reverberating bang echoed through the locker room. The sound was sharp and sudden, like something heavy had been hurled against one of the metal lockers and then clattered to the floor.

Bondsy, who was recording audio at the time, caught the entire exchange. The following is a transcript of the incident, minus the swear words the team instinctively let slip.

Transcript of the First Locker Room Bang:

Bondsy: "Can you take a flash burst with that camera?"

Chris: "Yeah."

Bondsy: "Then take a whole bunch, dude!"

Two seconds later, the loud bang occurs, followed by a sound eerily similar to the glass from earlier on the second floor.

Bondsy: "That is not funny!"

Chris: "That's not you, Trigger?"

Trigger: "No, I'm shaking!"

Trigger's voice trembled with fear, her usual bravado completely stripped away by the unknown force that had just made its presence felt. The tension in the room was evident and unfortunately for Trigger, this would not be the last unnerving event she would experience that night. Not long after, something disturbingly similar would happen again, deepening the mystery of the locker room.

Lil Beans and Chris decided to head back to the pool area, leaving Bondsy and Trigger to explore the dimly lit locker room. Nearby, there was an old sauna that Bill had mentioned earlier—a place where, according to local lore, a man had died under mysterious circumstances. Intrigued, Bondsy and Trigger decided to take turns sitting inside the sauna to see if they could stir up any paranormal activity.

At one point, while Bondsy was inside the sauna, he heard the distinct sound of the door handle jiggling. Startled, he called out to Trigger, asking if she was the one messing with the handle. Trigger, standing just outside the room, quickly denied it. The unease that had been slowly building inside Bondsy reached a peak, and he decided he'd had enough of the sauna. He exited, the oppressive atmosphere of the tiny room clinging to him like a second skin.

They continued their investigation, winding their way through the labyrinth of lockers and narrow corridors. Eventually, they found themselves back in front of the sauna. Bondsy, trying to lighten the mood, asked Trigger if she wanted to give it a shot and sit inside. Just as Trigger began to respond with a hesitant, "I guess," the silence was shattered by a loud bang. The sound echoed off the metal lockers, identical to the noise they had heard earlier— something small and hard, like a nail or a piece of glass, hitting metal and then bouncing to the floor.

Here's the transcript from Bondsy's audio recorder capturing the moment:

Transcript of the Second Locker Room Bang:

Bondsy: "Do you want to go in the sauna?"

Trigger: "I guess."

[Immediately after Trigger's response, the loud bang occurs, followed by a startled expletive from Bondsy and a nervous giggle from both.]

Bondsy: "That's the same damn thing!"

Trigger: [Laughs nervously.]

Bondsy: "I want to go back to the pool; is that the way back to the pool?"

Trigger: "I don't know!"

Trigger: "Here, this is the way back to the pool!"

With that, they quickly left the area, the atmosphere charged with nervous energy. Throughout the night, it seemed that things were happening around Bondsy. I began to wonder if this was somehow connected to the uneasy feeling he'd had earlier on the second floor. Objects had been thrown at or near our group three times that night, and on each occasion, Bondsy and Trigger were present for each episode. Whether they were the specific targets of these incidents or simply caught in the crossfire of something larger remained unclear. But the odds of being in the wrong place at the wrong time three times in one night felt like more than just coincidence.

As the night wore on and the adrenaline began to fade, Bondsy and Trigger met up with Chris and Lil Beans, and the foursome made their way back upstairs to where Sarah and I were wrapping things up. It was just after 3:30 a.m., and with Bondsy and Sarah needing to be back in Springfield for their 6:00 a.m. radio show, we reluctantly began packing up our equipment. Though we had to cut the investigation short, the night had been anything but uneventful—leaving us with more questions than answers and a night we would long remember.

In my years of investigating the supernatural, physical interaction with unseen forces is a rarity. While unexplained sounds, shadows, and strange occurrences are common, the tangible movement of objects—especially in ways that defy logical explanation—stands out as particularly extraordinary. In fact, the only experiences more uncommon have been the times I've witnessed actual apparitions—a mere seven occurrences in two decades of investigation. Whether the entities responsible for these phenomena are spirits or some other form of life force that we have yet to fully understand, remains one of the great mysteries in this field.

Investigation Summary

The activity we encountered at the YMCA took on a distinctly physical form. Objects were thrown, though whether they were aimed at us or simply near us is something we may never know. But one thing was certain: something was intent on making its presence known. I often wonder about the mechanism behind these occurrences. Were these objects propelled by an invisible hand, or was some other supernatural force at play, manipulating matter in ways we cannot yet comprehend? Whatever the cause, it happened not once, but three times that night—an undeniable pattern of interaction.

The persistence of these incidents led me to believe that whatever was there wanted us to be aware of its presence. But the question remains: why? Was it simply trying to communicate, or was it attempting to frighten us into leaving? The latter seems plausible, given that when we tried to engage through EVP sessions, the only discernible response came from the swimming pool area, where Sarah's questions directed at the spirit of a little girl elicited not just a vocal response, but also a physical sensation—a touch on her leg that startled her in a way no mere coincidence could.

Most of the interactions that night were unprovoked, at least from our perspective. Take the shushing sound we heard and later recorded by the pool. It came out of nowhere, without any immediate trigger on our part, as though some unseen entity was reminding us to keep our voices down. The sound was eerily reminiscent of a stern librarian silencing a noisy student or a teacher calling for order in a chaotic classroom. Perhaps our mere presence, our voices echoing through those empty spaces, was all the provocation needed to stir whatever lay dormant there.

What remains in the aftermath of that night are questions that may never be fully answered. Were these events the result of restless spirits trapped in a place that once bustled with life, or were we dealing with something far more complex—a force that we have yet to name or understand? What I do know is that nights like this one keep me coming back, driven by the need to unravel the mystery of what lies beyond the veil of our everyday reality.

The old YMCA is an expansive, convoluted building, and our night of investigation was marked by activity that seemed to follow us throughout the entire facility. This raises the question: were we dealing with multiple spirits, each tied to a specific part of the building, or was there one particularly restless phantom shadowing our every move? My instincts tell me it was the former. There's a sense that more than one ghostly entity calls this place home, each contributing to the tapestry of strange occurrences we encountered.

One thing stood out during our investigation: the curious connection between the thrown objects and the presence of Bondsy and Trigger. They were there every time something was hurled through the air, almost as if they were magnets for whatever was trying to contact us. Was it mere coincidence, or were they unwittingly attracting the entity's attention? Ian Fleming once wrote, "Once is happenstance, twice is coincidence, and three times is enemy action." By that logic, it's hard to dismiss the idea that something was deliberately targeting them—or at least making sure they were in the thick of it.

Bondsy's discomfort was apparent, especially when we ventured to the second floor. It was as if something was deliberately trying to unsettle him, testing the boundaries of his resolve. Or perhaps it was that primal sixth sense, the one that warns us of danger even when our rational minds can't pinpoint the source. Whatever the reason, Bondsy's unease seemed justified. The second floor, particularly near the stairwell, has an undeniable presence. It's a place where the air feels thicker, and where every creak and groan of the old building seems amplified by an unseen force.

The drastic temperature drop we recorded on that very floor was a physical manifestation of this eerie atmosphere. In a matter of seconds, the temperature plunged from a stifling eighty degrees to a bone-chilling twenty-two. Such a sudden, dramatic change defies normal explanation and adds weight to the theory that something otherworldly is at play.

So, did our experiences on the night of July 25, 2013, provide proof of the paranormal? If disembodied voices captured on tape, objects thrown by invisible hands, and a sixty-degree temperature drop in mere seconds qualify as evidence, then the answer is an unequivocal "yes." I believe the old Granite City YMCA is haunted, and I suspect it will remain a hub of paranormal activity for as long as the building stands.

If you ever find yourself lucky—or unlucky—enough to investigate the old YMCA in Granite City, Illinois, you might want to bring along a safety helmet. It could come in handy if the spirits roaming those darkened halls decide to make their presence known in the same way they did for us—by hurling a few more objects into the air.

SOMETHING SPECIAL FLORIST
VANDALIA, ILLINOIS

THREE

On February 25, 2012, I found myself standing outside a pre–Civil War era building at 402 West Gallatin Street in Vandalia, Illinois, ready to investigate what many locals had come to regard as a paranormal hotspot. The structure, with its weathered bricks and aged windows, loomed large in the chilly evening air—a silent witness to over a century and a half of history. The building had seen better days, but its reputation for unexplained occurrences had only grown stronger over time.

Joining me in this investigation was a fellow paranormal investigator, whom I'll refer to as Mr. Smith. Due to the sensitive nature of the events, he would later experience, I decided it was best to keep his name anonymous. Along with Mr. Smith, I was accompanied by Sarah Hunter, the charismatic co-host of the morning show on 99.7 KISS FM, and her friend Erica Lindsey. Unfortunately, Bondsy, my usual partner in these ventures, was unable to join us due to a family commitment. His absence was a disappointment, but it also heightened the sense of anticipation—there was something about this night that felt different, almost charged.

The building itself was a mix of the past and present. The first floor housed several businesses, each with its own unique charm: a quaint gift shop, a wine store, an H&R Block tax service, and a flower shop where the scent of fresh blooms mingled with the musty odor of age. But it was the upper floors that intrigued me the most. These spaces, largely untouched by modern hands, were used primarily for storage—a repository for forgotten relics and, perhaps, if the stories were true, lingering spirits.

In recent years, employees working within these walls had reported encounters that defied explanation. The most common sightings involved two distinct figures: the apparition of a large man and that of a little boy, often seen peeking around the corner of a workbench in the flower shop. The staff had come to know these spirits well—or as well as one can know an entity that exists on the other side of the veil. The man, they believed, was a former tenant named Bill Laswell, a name spoken with equal parts reverence and trepidation. He was a constant presence, his apparition seen during both day and night, while the boy's appearances were more fleeting, his small form vanishing almost as soon as it was noticed.

As we prepared to enter the building, I couldn't shake the feeling that we were about to step into a place where the past had never fully let go. This wasn't just another investigation; it was a journey into the unknown, where history and the paranormal intertwined in ways that would soon challenge our understanding of the world around us.

This was my third investigation of the building at 402 West

Gallatin Street, and it would prove to be the most eventful yet. The old structure, with its creaking floors and shadow-filled corners, had never failed to stir the imagination, but tonight would push the boundaries of what we believed possible.

Our primary focus this night was a large room on the second floor, a space that had long since lost its original purpose but not its connection to the past. Within this room was a doorway that led to a smaller, more intimate space—a room that had once been home to Bill Laswell. Bill's living quarters were modest, almost solitary in their simplicity. A hot plate on the counter served as his kitchen, and a small bathroom with a sink and toilet, now covered in a thick layer of dust, offered the barest of comforts. It was a forgotten room, but it carried the weight of memories, both known and hidden.

For some time, we staked out the large room, sitting in silence as we surveyed the surroundings. From our vantage point, we could see not only into Bill's old room but also down the hallway that stretched beyond. The atmosphere was heavy, as if the very air was thick with anticipation. We all felt it—something was going to happen.

Around 1:30 a.m., we decided to change our approach. Moving our chairs into the hallway, we set up for an (electronic voice phenomenon), EVP session and a spirit box session, hoping to contact whatever entities might be lingering in the shadows. I positioned myself about eight feet down the hall, facing the doorway to the large room. Erica sat to my left, Sarah to my right, while Mr. Smith leaned casually against the wall, his back to the large room but his senses on high alert.

We hadn't been in our new positions for long when the first sign of something unusual occurred. A flash of light, brief but unmistakable, flickered from within the large room. I immediately spoke up, alerting the group to what I had seen. My words were barely out of my mouth when Sarah confirmed my sighting, her voice tinged with both excitement and apprehension.

"I saw it too!" she exclaimed, her eyes wide with astonishment.

Mr. Smith, ever the skeptic, saw the light as a reflection on the staircase in front of him that led to the third floor. But his curiosity was piqued, nonetheless. Without hesitation, he headed into the room to investigate. As soon as he crossed the threshold, he stopped in his tracks.

"It's freezing in here," he announced, his breath visible in the sudden chill.

I quickly followed him, and the temperature drop was immediately apparent. The room, which had been merely cool before, was now icy cold, as if the warmth had been sucked out of it. I called Sarah and Erica to join us so they could feel the difference for themselves.

As soon as Sarah entered the room, she let out a startled cry, her eyes locking onto the doorway of Bill's old room. "I just saw a shadow move!" she shouted, pointing toward the small, darkened space.

My heart raced as I asked her what exactly she had seen. Her response sent a shiver down my spine.

"I saw a large shadowy figure going into Bill's old room," she said, her voice shaking slightly.

A large shadowy figure. The words hung in the air like a challenge. In that moment, I knew exactly who—or what—she had seen. Bill Laswell. His presence, felt so strongly by those who had encountered him in life, seemed to be just as potent in death.

Without thinking, I moved toward Bill's room, my pulse quickening with every step. I wasn't sure what I expected to find—a ghostly apparition, perhaps, or maybe just the lingering sense of something watching us from the shadows. But whatever it was, I was determined to confront it, to get a glimpse of the past that refused to stay buried in this old building.

As I crossed the threshold into Bill's room, the air grew even colder, and the shadows seemed to deepen around me. For a moment, everything was still—eerily, unnaturally still. Then, just as suddenly as the light had flashed and the shadow had moved, the room was plunged into a silence so profound that it seemed to muffle even the sound of my own heartbeat.

Whatever had been there moments before was gone, leaving only the cold and the darkness behind. But the experience had left its mark. We had come looking for signs of the paranormal, and the building had delivered—just not in the way any of us had anticipated.

As Erica stepped into the large room, joining Smith and Sarah, the atmosphere seemed to thicken with a detectable tension. The energy in the room was unlike anything I'd felt before, an almost electric charge that raised the hairs on the back of my neck. After a brief exploration of Bill's old quarters, I turned to rejoin the rest of the team in the larger space.

With each step I took, I could feel the weight of the night pressing down on me, an unspoken anticipation hanging in the air. Smith stood near the doorway leading to the hallway, his digital camera in hand. The camera he used emitted a pre-flash—a quick strobe effect that momentarily illuminated the room like a series of miniature lightning strikes. It was through these brief bursts of light that I navigated my way back into the room, the litany of flashes offering just enough illumination to see where I was going.

As Smith prepared to take another shot, I felt a sudden shift in the air, a subtle change that hinted at something just beyond the veil of the ordinary. And then, as the next flash of light cut through the darkness, I saw them.

Standing no more than six or seven feet in front of me were two small children, their forms unmistakable in the brief but brilliant light. It was as if they had materialized from the very shadows, their presence both startling and surreal. My heart skipped a beat, and without thinking, I yelled out, "I see two kids!" My voice echoed through the room, filled with a mix of awe and urgency.

I pointed to the spot where they stood, urging Smith to keep taking pictures, convinced that the strobe light effect from his camera was somehow making the children visible to us. Smith obliged, his camera flashing again in rapid succession. The room was momentarily bathed in light, but this time, the children were gone.

Before I could process their disappearance, Sarah cried out, her voice tinged with both excitement and fear. "I see them! They're holding hands and running toward the door!" She pointed frantically in the direction where the children had supposedly fled, but when I turned to look, there was nothing—just the empty, shadow-filled room.

Then, abruptly, the camera flashes ceased. The room was plunged back into darkness, and a heavy silence fell over us. I turned toward Smith, expecting him to be frantically clicking away, trying to capture any trace of the apparitions we had just witnessed. Instead, I saw something that sent a chill down my spine. Smith's back was turned, and he was slowly walking out of the room.

A wave of unease washed over me. This was not like Smith. He shared my passion for the supernatural, and in all our years of investigating together, I had never seen him react this way. When faced with the unexplained—especially something as rare as an apparition—Smith would usually be the last person to turn his back and walk away. Something was off, deeply off, and I could feel it in the pit of my stomach.

As I watched him leave the room without a word, a nagging suspicion took root in my mind. Had he seen something that shook him to his core, something he couldn't explain? Or was there another force at play here, something that was influencing his actions in ways we didn't yet understand?

I shouted after him, "Where are you going?" My voice echoed through the darkened room, but Smith didn't stop. Instead, he responded in a voice that was unnervingly slow and deliberate, a voice that didn't sound like his own. "I feel sad," he said, each word dripping with an eerie finality.

His reply sent a shiver down my spine. "You feel sad? Sad about what?" I asked, trying to make sense of the situation. But he didn't answer. Smith kept walking, his steps mechanical, as if something— or someone—was guiding him. He headed straight for the wall of the stairwell that led to the third floor, moving with a strange, trance-like purpose.

Once he reached the wall, he raised his arm and pressed his forehead against it, leaning heavily as if the weight of the world had suddenly come crashing down on him. He stood there, unmoving, his posture rigid and unnerving.

Alarm bells were ringing in my head. This wasn't the Smith I knew. The Smith I knew was level-headed, analytical, and unshakable, even in the face of the unexplained. I moved quickly toward him, desperate to find out what was wrong.

It wasn't until I got closer that I realized what had happened after Smith had taken that last photo. Without a word, he had shoved his camera into Sarah's stomach, almost as if the device had become an unbearable burden. Sarah, despite being relatively new to paranormal investigating, had sensed immediately that something was very wrong. But instead of panicking, she took control, continuing to snap photos with Smith's camera, her hands steady despite the growing tension in the room.

Sarah had always been curious about the paranormal, driven by a deep-seated need to know if the stories were true. That curiosity seemed to fuel her courage, allowing her to push past any fears she might have had.

As I passed through the doorway and approached Smith, I placed a hand on his right shoulder and gently turned him to face me. The beam from my flashlight cut through the darkness and illuminated his face—and what I saw sent a jolt of fear straight through me.

Tears were streaming down Smith's cheeks, glistening in the cold light. But it was his eyes that truly unnerved me. His pupils had rolled up into his head, leaving only the whites visible, giving him an almost otherworldly appearance. It was as if the person I knew had been replaced by something else, something that wasn't entirely human.

"Smith, what's wrong, buddy?" I asked, my voice laced with concern. I gave him a gentle shake, hoping to snap him out of whatever trance he had fallen into. But his response was the same as before—slow, deliberate, and disturbingly detached.

"I feel sad," he repeated, the words seeming to hang in the air like a dark cloud. This wasn't just sadness. This was something deeper, more insidious, something that had taken hold of him and wasn't letting go.

I felt a rising sense of dread. This was beyond anything I had encountered in my years of investigating the paranormal. Something had connected with Smith, something that wasn't content with merely making its presence known. It wanted more—it wanted to take control.

As I stood there, gripping Smith's shoulders, trying to pull him back to reality, I couldn't shake the feeling that we had stumbled upon something far darker than we had anticipated. And whatever it was, it might not be done with us yet.

He started to sway, his legs buckling as if the strength were draining from his body. I grabbed him by the shoulders, trying to steady him before he collapsed. "Let's get you over to a chair so you can sit down," I said, trying to keep my voice calm despite the growing panic gnawing at me.

But Smith was having none of it. "I don't want to sit down," he kept insisting, his voice laced with a fear that was now discernible. His eyes, still rolled up into his head, flicked around the room as if searching for something—or someone. He twisted around repeatedly, glancing over his shoulder with a look of sheer terror.

"Who's behind me?" he asked, his tone trembling. "There's someone behind me."

I quickly shone my flashlight into the darkness behind him, illuminating the empty space. "There's no one there, Smith," I reassured him, trying to break through the fear that had taken hold. But it was no use. He was convinced that something—or someone—was lurking just out of sight.

Despite his protests, I knew I had to get him seated, to ground him somehow. His eyes were still rolled back, and I could see he was teetering on the edge of unconsciousness. I kept talking to him, trying to keep him focused, to keep him anchored in the present moment.

After what felt like an eternity, I finally managed to guide him to one of the chairs in the hallway. But even then, it wasn't easy. I had to practically force him to sit, pressing down on his shoulders to keep him in the chair. As soon as he was seated, his demeanor shifted again, this time from fear to anger. His expression hardened, and he began to radiate frustration and agitation.

Erica, sensing that something was seriously wrong, came over to see if she could help. She spoke to Smith, her voice soft with concern, trying to reach out to him, to comfort him. But as soon as she spoke, Smith turned on her with a look that I'll never forget— an almost crazed glare that made it clear he was ready to lash out. The shift in his emotions was startling, from sadness to confusion, and now to raw, unbridled anger.

In those few short minutes, Smith's personality had twisted and turned through a spectrum of intense emotions. It was as if he were being pulled through a storm of feelings that weren't entirely his own. Erica took a cautious step back, sensing the danger that simmered just beneath the surface.

I stayed close, keeping a firm grip on his shoulders, trying to steady him, both physically and emotionally. After what seemed like an eternity, the storm inside him began to subside. His breathing slowed, and the wild look in his eyes started to fade. Gradually, he began to snap out of it, the tension in his body easing as he started to regain his senses.

The entire episode lasted no more than five minutes, but those were the longest, most intense five minutes of the night. In that short span of time, I found myself genuinely worried not just for Smith's safety, but also for Erica and Sarah. Whatever had taken hold of him was powerful, unpredictable, and more than a little terrifying.

So, what exactly happened to Smith that night? Was his body taken over by spirits? The term "possession" might be too strong, or at least not applicable in the traditional sense. But something profound and unsettling did occur, something that defies easy explanation.

Let me walk you through what I believe transpired. Before Smith's strange behavior began, he was standing in the doorway between the large room and the hallway, snapping photos with his camera. I had just exited Bill Laswell's old room and was moving toward the center of the large room when Smith's camera flashed.

In that brief, blinding moment, I saw them—two children, a boy and a girl, standing no more than a few feet away. They were dressed in old-fashioned clothing, the kind you might associate with the days of *Tom Sawyer* and *Huckleberry Finn*. The building we were investigating dated back to the mid-1800s, so it was likely these children were from that era.

The boy stood to my left, the girl just to his right. They weren't moving, and their faces were eerily expressionless. They looked to be around eight or nine years old. I stared at them, and they, in turn, were staring directly at Smith, who was still in the doorway. What struck me was how they seemed like something out of an old horror film—their faces blank, their arms hanging limply at their sides, and everything about them devoid of color, as if I were looking at an ancient black-and-white photograph.

But there was one exception: the little boy's chest bore a pale-yellow splotch, a faint hint of color that stood out starkly against the rest of his monochrome appearance. It looked like it might have been the original color of his shirt, and I could see that his sleeves were either rolled up or designed to be three-quarter length.

Despite the eerie nature of the encounter, something told me these two were siblings. There was a bond between them that transcended their ghostly state. But what truly unnerved me were their eyes—coal black, empty, like the eyes of the so-called "black-eyed kids" you sometimes hear about in urban legends. Those eyes were unsettling, to say the least, and they were fixed intently on Smith.

I believe these children had been following us throughout the night, thinking they were invisible to us. And they were, until that camera flash. Somehow, the combination of the strobe effect and the camera's flash had temporarily pierced the veil that usually hides such entities from our sight. It was as if the light had momentarily slowed down, filtered, or distorted the spectrum, allowing us to catch a glimpse of these spectral siblings.

When I shouted, "I see two kids!" it must have shocked them as much as it did me. Realizing they had been seen, they panicked, just as any frightened children might, and bolted for the nearest exit. And that exit, unfortunately, was the very doorway where Smith stood.

Here's where things take a darker turn. In their frantic rush to escape, one or both of those children seemed to have passed right through Smith. Now, imagine what that must feel like—an ethereal being, a remnant of a time long past, surging through your body. The trauma of such an experience is hard to fathom.

Whatever the effect on Smith, it was severe. The man was shaken to his core, though mercifully, he remembers very little of what transpired. But the brief flashes of memory he does have, combined with our observations, suggest that the encounter left a profound mark on him—a chilling reminder of the unknown forces we were dealing with that night.

Many paranormal investigators subscribe to the idea that ghosts or spirits are essentially forms of energy. While I agree that spirits can be perceived as energy, I contend that if they possess awareness of their surroundings, they are far more complex than mere energy. They are entities with thoughts, emotions, and a form of consciousness that transcends simple energy.

Consider Smith's case. If, as it appeared, one or both of the children's spirits passed through his body, what kind of biological and psychological impact might that have had on him? The transformation we witnessed in Smith was profound—his demeanor, personality, and emotions shifted dramatically within a matter of minutes. It was as if he were momentarily possessed by the very spirits he had encountered.

Did Smith, in those fleeting moments, inadvertently absorb or take on the thoughts and emotions of the children? The rapid change in his behavior suggested as much. Whether intentional or not, this experience proved to be a harrowing ordeal for him and the rest of us. It was a reminder of the uncontrollable and unpredictable nature of supernatural encounters.

I want to make it clear that I am not attempting to deter anyone interested in the field of paranormal investigation. However, it is crucial to recognize the inherent risks involved. The realm of the supernatural is fraught with uncertainties and forces that we do not fully understand or control.

Since the incident with Smith, I have adopted a new precaution: I no longer position myself in doorways, entrances, or exits during investigations. The lesson learned was stark and undeniable—paranormal investigation is a captivating and thrilling pursuit, but it must be approached with the utmost respect and caution. While careful and respectful investigation can help mitigate risks, neglecting these principles can lead to unforeseen and potentially dangerous consequences.

Investigation Summary

The February 2012 investigation of the pre-Civil War building in Vandalia, Illinois, proved to be far more than we had anticipated. Initially, we approached this site with a mix of skepticism and curiosity, armed with tales of ghostly encounters reported by employees—stories of the ghost of Bill Laswell and a little boy seen peeking around the counter of the flower shop. Yet the reality that unfolded was an encounter with the unexplained that would leave a lasting impression on each of us.

The night began with an uneasy stillness that permeated the air and unfolded when Sarah caught sight of a dark, shadowy figure in Bill Laswell's old room on the second floor. I couldn't help but wonder: Was she witnessing a remnant of Bill's spirit, perhaps engaged in the mundane tasks of his former life, or was this an entirely different entity, suggesting layers of history entwined with the living?

But it was the appearance of two ghostly children that truly rocked the foundation of our understanding. Both appeared suddenly, illuminated by the flash of Mr. Smith's camera, and their presence was like stepping into a Mark Twain novel. They stood side by side, their grayscale figures a stark contrast against the backdrop of the dilapidated building. Their eyes, as black as coal, held a haunting awareness as they seemingly watched Smith in intrigue as he stood in the doorway taking photos.

I urged Smith to take more photos, hoping to witness the spectral figures again, but I was disappointed when I didn't see them. Then, in a sudden burst of excitement, Sarah yelled out, "I see them! They're holding hands and running toward the door!" And just like that, the atmosphere shifted from eerie stillness to frenzied chaos as the children dashed toward the very doorway Smith occupied.

What unfolded next felt scripted—like a scene straight out of a Hollywood horror movie. The two spectral children, caught in a moment of panic, exposed by the camera flash, rushed toward Smith, their celestial forms moving through him as if he were air. This interaction left Smith profoundly affected, a receptacle for the emotional energy they discharged. As I watched, it became painfully clear: We were no longer mere spectators but active participants in something we could barely comprehend.

When I yelled, "There are two kids there!" I sensed we had crossed a threshold. The children, having been aware of our presence, felt the pressure of being seen and reacted instinctively. In that moment, Smith became a conduit, absorbing their emotional turmoil—sadness, fear, and anger. The transformation in him was palpable, a testament to the weight of the encounter.

After the dust settled, Sarah approached me, grappling with the extraordinary nature of what we had just experienced.

"Maybe by saying that you saw the ghost kids, you put that thought in my mind, and I only thought I saw what I saw," she pondered, her skepticism fighting with the undeniable reality of the moment.

"Well, tell me what you think you saw, and then I'll tell you what I saw," I replied. "We wouldn't both hallucinate the same thing."

"No, you tell me first," she answered back, her skepticism revealing itself again.

"What I witnessed when Smith's camera flashed was a young boy and girl, ages eight to ten, dressed in clothes reminiscent of the Tom Sawyer era. The boy had dark, shoulder-length, Dutch boy-style hair and wore a shirt with either three-quarter sleeves or rolled-up sleeves. The only color was a yellow splotch on the boy's chest. The girl was in a long white dress, had shoulder-length curly hair, and both figures were completely devoid of color, like an old black-

and-white photo. The most chilling thing was their coal-black eyes," which reminded me of drawings of Black-Eyed Children I had read about."

Sarah's expression changed; her mouth dropped in disbelief. "Larry, that is exactly what I saw!"

In that exchange, it was clear we had both encountered the same ghostly figures, validating not only our experiences but also the reality that we had brushed against something beyond the veil. This investigation left us with more questions than answers and a deeper respect for the unknown.

As we left the building, the weight of what we had witnessed hung in the air. The ghostly remnants of the past lingered, an echo of lives lived long ago, reminding us that the boundary between our world and the next is much closer than we realize. The night's revelations served as a cautionary tale—one that I urge future investigators to approach with both wonder and respect, for the world of the paranormal is as unpredictable as it is compelling.

FARRAR SCHOOL
FARRAR, IOWA

FOUR

No more pencils, no more books,
No more teachers' dirty looks
School's out for summer,
School's out till fall,
We might not come back at all.

The above lyrics from Alice Cooper's 1972 hit "School's Out" may sum up the sentiments of students eagerly fleeing the confines of their classrooms for summer vacation, but they might not apply to the children who once attended Farrar School in Farrar, Iowa. In fact, if the accounts of paranormal activity are to be believed, the children never actually left.

The old schoolhouse, now long abandoned, is said to be haunted by the spirits of its former students. Eyewitness reports tell of ghostly children standing on the stairs, their forms flickering in the dim light, and of locker doors slamming shut on their own, as if some unseen force is still keeping the school's routines alive. It's as if the bells are still ringing, summoning students to class—but this time, class is in session 24 hours a day, and no one is going home.

During our investigation of Farrar School, I took on an unusual role: that of a substitute math teacher. In an impromptu attempt to connect with whatever or whoever was still lingering in the halls, I decided to test the waters with a math quiz. To my surprise, someone—or something—responded. Even more astonishing, the answers given were correct. And the most remarkable part? We didn't just hear the responses; we recorded them. You'll read about this eerie encounter in detail shortly.

Over the years, there have been numerous reports of a tall, shadowy figure seen in various locations within the building. Dubbed the "Shadow Man," this three-dimensional apparition is said to stand seven feet tall, a towering presence that seems more fiction than reality—until you see him for yourself.

Schools are generally considered safe and welcoming places, filled with the laughter and energy of children. But when those buildings are left empty, a strange transformation occurs. The atmosphere shifts, and the once-vibrant classrooms and hallways take on an unsettling vibe, as if the echoes of the past have taken over. It's a change that's hard to describe but easy to feel—like stepping into one of R. L. Stine's *Goosebumps* books, where familiar places become the settings for nightmares.

Farrar School, with its lingering spirits and strange phenomena, is a stark reminder of how places we think we know can be anything but. That there is much more going on than we can merely see with our eyes. The children may have been dismissed for summer break long ago, but it seems they've never really left. Instead, they remain, their restless souls keeping the lessons going long after the final bell has rung.

A Brief History of Farrar School

Farrar is a small, unincorporated community nestled in Polk County, Iowa, a place that time seems to have forgotten. With just a handful of homes, a church, a cemetery, and the imposing 17,000-square-foot school building, Farrar feels like a snapshot of rural America on the decline. Like many small towns across the country, its population has dwindled over the years, and the once-vibrant heart of the community—the school—now stands as a silent sentinel to a bygone era.

The history of Farrar School is rooted in the early 20th century, a time when the country was on the cusp of modernization. According to the school's official website, the land on which the school now stands was donated in 1919 by a local farmer, C. G. Geddes. The vision was ambitious: to consolidate all the one-room country schoolhouses scattered across the area into one central, modern facility. This new school would be known as the Washington Township Consolidated School District.

Construction of Farrar School was completed, and on April 1, 1922, the community gathered for a grand dedication ceremony. However, the celebration was not without controversy. The new building, with its $100,000 price tag, was considered extravagant by some locals. A boiler heating system, electric lights, and indoor plumbing were seen as unnecessary luxuries in a schoolhouse, leading some to dismiss the project as "a monument to the arrogance and vanity of the school board." Despite these criticisms, the school became a cornerstone of the community, educating generations of children.

Fast forward to May 3, 2002—after eighty years of service, the school closed its doors for good. The building sat abandoned for four years until Jim and Nancy Oliver purchased it in December 2006. The Olivers, who are in the process of restoring the historic structure, now live in a portion of the building that serves as their home. But it didn't take long for them to realize they weren't the only ones residing there.

During a walkthrough of the building prior to our investigation, Nancy Oliver recounted several unnerving incidents. One of the most memorable occurred on a staircase when she lost her balance and nearly fell. She distinctly felt a hand on her shoulder, steadying her and preventing a potentially serious accident. Assuming it was her husband, she turned to thank him, only to find she was completely alone.

Nancy's story is just one of many. A dark, distinct outline of a small boy has been seen on the stairway leading down to the gymnasium. Witnesses describe the figure as about three and a half feet tall, standing on the stairs and holding onto the handrail. The boy remains motionless for several seconds before fading away, as if he's simply returning to wherever these spirits go when they're not making themselves known.

Over the years, both students and staff at Farrar School have reported a litany of strange occurrences—disembodied voices echoing through the empty halls, unexplained noises that seem to come from nowhere, and the unsettling appearance of apparitions.

Psychics who have visited the site claim that the spirits of the dead have taken up permanent residence in the former schoolhouse. Paranormal investigators, too, have documented unusual activity, adding to the building's growing reputation as a hotspot for the supernatural.

I first heard about Farrar School in 2011, on one of the many television shows dedicated to ghosts and hauntings. The idea of investigating a former school piqued my interest immediately. I'd explored many types of buildings in my career, but an old schoolhouse was something new—a place where the echoes of children's laughter might still linger in the air, mixed with the whisper of things unseen. The prospect of delving into the mysteries of this storied location was irresistible, and I knew I had to experience it for myself.

When I decided to investigate Farrar, I knew it would be the perfect opportunity for my friends Bondsy and Sarah, who were scouting locations for their 2013 Halloween show. A sprawling, eerie school building supposedly haunted by the spirits of children—what could be better? As soon as I mentioned the possibility of exploring such a place, Bondsy was on board, his enthusiasm matching mine. We quickly assembled our team for the trip to Farrar, which included not only Bondsy and Sarah but also Chris, the investigator who had joined us during our investigation of the Granite City YMCA in Illinois.

The deeper I venture into the world of the supernatural, the more attuned I become to the peculiar things that seem to happen just before an investigation. Some might dismiss these events as mere coincidence, but I've come to see them as synchronistic or, at the very least, unusual enough to take notice. They're often subtle—not the sort of thing that grabs headlines—but they have a way of making me pause, scratch my head, and wonder, "What the heck is going on?"

The day of the Farrar investigation was no different. It began with a couple of strange incidents that set the tone for what was to come. The first occurred at a local gas station, where I stopped to fill up my SUV before picking up the team. As I stood by the pump, waiting for the tank to fill, I wasn't thinking about anything out of the ordinary. But then, as the pump clicked off, I glanced at the display and saw that I had put exactly 11.11 gallons of gas into the tank.

For most people, that might have been nothing more than a curious little quirk, a numerical oddity with no real significance. But for me, it was impossible to ignore. I've had too many experiences with the 11:11 phenomenon to simply brush it off. Those four numbers staring back at me from the pump immediately caught my attention. It wasn't earth-shattering, but it was enough to make me feel like I was once again stepping into the realm of the unknown— like something out there knew exactly what I was about to do and wanted to make sure I knew it too.

It was a subtle reminder, a small nudge that said, "We're watching." And as I climbed back into the SUV, ready to embark on another journey into the supernatural, I couldn't help but feel that this investigation at Farrar School was going to be anything but ordinary.

After the incident with the gas pump, I picked up the team, and we hit the road, heading toward Iowa. The journey had been uneventful, just the usual banter and anticipation building as we got closer to Farrar. We had been driving for about three hours when Sarah suddenly realized she'd forgotten to bring the charger for her cell phone. Not wanting to risk running out of battery during the investigation, we decided to pull off the highway at the next exit, which led us to a large truck stop. This is where the second of the day's unusual pre-investigation occurrences took place.

We parked near the front entrance of the store. Bondsy and Chris were the first to get out, eager to stretch their legs and grab some snacks. I waited for Sarah to gather her things so I could lock the SUV, and then we followed behind the others. Inside the store, I noticed Chris already browsing one of the aisles. Bondsy was heading toward the soda fountain area, but I soon lost sight of him as he disappeared behind a display. Sarah was nearby, scanning the shelves for a phone charger. It was then that I noticed a man walking out of the men's restroom, the door closing behind him. I figured it was as good a time as any to make a quick pit stop before we got back on the road, so I headed that way.

The restroom door was one of those simple designs, with no doorknob, just a shiny metal plate where a handle might have been. As I approached, I reached out with both hands to push the door open, but to my surprise, the door didn't budge. My momentum caused me to bounce backward slightly, confused by the unexpected resistance. The door was locked. I gave it another push, this time with more force, but it still wouldn't open.

I stood there for a moment, perplexed. I hadn't seen anyone enter the restroom after the man who'd just walked out, so the door shouldn't have been locked. I scanned the store, trying to figure out if Bondsy had somehow slipped past me and locked himself in there, but I quickly spotted him still lingering by the soda fountain on the far side of the store. So, it couldn't have been him. Still puzzled, I decided to wait outside the restroom, casually looking at some nearby merchandise.

A minute later, Chris came strolling toward me. Without hesitation, he pushed the bathroom door with both hands, just as I had done, and this time, it swung open effortlessly. My jaw dropped. I couldn't believe what I was seeing—seconds ago, that door had been locked tight. I glanced over at the soda fountain again, and there was Bondsy, still fiddling with his drink. He hadn't been near the restroom at all.

Feeling a mix of confusion and curiosity, I walked over to the now-open restroom door and gave it another push. It opened easily. I stepped inside and found Chris alone in the room. I asked him if Bondsy had been in there before him, but Chris shook his head, saying the restroom was empty when he walked in.

I explained what had just happened—how the door had been locked when I tried to enter, and how it was inexplicably unlocked moments later. Chris reassured me that there was no one else in the restroom before he arrived, and the door had opened without issue for him. We both looked around the small room, confirming that there was only one way in and out. I examined the door again, but there was no lock on it—just a simple metal plate. There was no logical explanation for why the door wouldn't open for me.

As we left the store, I mentioned the strange occurrence to Bondsy. He confirmed that he hadn't been anywhere near the restroom. The incident gnawed at me as we continued our journey. It wasn't earth-shattering, but it was another one of those odd, unexplainable moments that seemed to happen more often than not when I am about to delve into the paranormal. I couldn't shake the feeling that it was another subtle sign that something—or someone—was aware of what we were about to do and was making its presence known.

Farrar is a small, unassuming dot on the map, situated about twenty-five miles northeast of Des Moines in the heart of rural Iowa. To get there, you turn off Interstate 80 and wind your way through several narrow country roads, each one seemingly more remote than the last. By the time we arrived at the old schoolhouse, it was around 8:00 p.m., and the weather had taken a dramatic turn. Lightning flickered on the horizon, casting brief but intense flashes of light across the surrounding fields—a fitting backdrop for what promised to be an evening drenched in the unknown.

As we pulled up to the school, its imposing structure loomed against the darkening sky. It was a relic of another time, its walls steeped in the memories of countless students who had passed through its halls. The building, though aged and weathered, still held a certain gravity, as if it had been waiting for us.

Soon after we arrived, we were met by Nancy Oliver, the current owner of the school. After brief introductions, she graciously offered to give us a tour, guiding us through the dimly lit corridors and empty classrooms that had once been filled with the sounds of children. But it wasn't just the echoes of the past that lingered here; according to Nancy, there was something far more tangible—and far more unsettling—still occupying these spaces.

As we made our way through the building, Nancy recounted one of her own experiences. It happened on the stairwell leading down to the gymnasium, a well-lit area that, under normal circumstances, wouldn't be considered particularly eerie. But what she saw there defied normalcy. A small boy, perhaps three or four feet tall, had appeared on the stairs, standing motionless with one foot on each step, his hand gripping the railing as if he were preparing to continue his descent. For a few seconds, he simply stood there, and then, just as suddenly as he had appeared, he vanished into thin air.

The matter-of-fact way Nancy described the encounter only added to its eeriness. She had seen this apparition with her own eyes—there was no mistaking what she had witnessed. This wasn't a story passed down or an anecdote overheard; it was something she had experienced firsthand.

Curious about the possible origins of the hauntings, I asked Nancy if anything tragic had occurred at the school that might explain the presence of these spirits. Her response was both disturbing and revealing. According to local rumors, there had been a janitor who worked at the school, a man who was suspected of molesting children during his time there. The unsettling part was that the school's principal at the time had allegedly been aware of the janitor's actions and may have even helped to cover them up.

Nancy knew the name of the janitor, but she was reluctant to share it, explaining that the man's family still lived in the area. The weight of that unspoken name hung in the air, adding to the unease of the already charged atmosphere of the school. As we continued our tour, it became increasingly clear that whatever was haunting this place was tied to more than just the memories of happy school days. There were darker stories lurking in the shadows of Farrar— stories that perhaps were never meant to see the light of day.

This was no ordinary school, and as the lightning continued to flash outside, casting eerie shadows across the walls, I couldn't shake the feeling that we were about to uncover something that had been waiting for a long time to be found.

Nancy had a way of telling stories that made even the most skeptical among us pause and reconsider what we thought we knew about the paranormal. One of the stories she shared with us that evening seemed too incredible to believe, yet it would soon become a reality for me in a way I hadn't anticipated. In just a few hours, I would come face-to-face, although through a video monitor with the very phenomenon she described—a phenomenon that I had only heard about in whispers and rumors until that night.

As we gathered in the dimly lit gymnasium, Nancy began

recounting an experience that had occurred a few years prior. A group of paranormal enthusiasts had visited the school for an investigation, and among them was a man who was not just any investigator—he was also a talented artist. During their exploration of the building, he encountered something extraordinary, something that, had he not seen with his own eyes, he might never have believed.

Nancy explained that the man witnessed a sight so incredible it left an indelible mark on his memory. He had seen the solid, dark silhouette of a figure moving through the school—a figure unlike any human he had ever encountered. What made this sighting all the more remarkable was the sheer size of the entity. It wasn't just tall; it was towering, at least seven feet in height, with an impossibly thin frame. The figure moved with a deliberate, almost mechanical grace, its arms and legs swinging back and forth as it walked through the halls of the old school.

Being an artist, the man had immediately sketched what he saw, capturing the entity in haunting detail. Nancy showed us the drawing, and as I studied it, I could see the precision in every line and shadow. The artist had captured the essence of what he had witnessed—this was not just a rough sketch; it was a memory etched into his mind as much as it was onto the paper. The figure he described was what paranormal investigators often refer to as a "shadow person," a type of entity that is as mysterious as it is unsettling.

As Nancy shared this story, I found myself wrestling with my own skepticism. I've always approached investigations with an open mind, but the idea of a seven-foot-tall shadow figure seemed almost too fantastic to take seriously. Yet there was something about the drawing and the way Nancy recounted the artist's experience that made it hard to dismiss outright.

After showing us the sketch, Nancy led us to the gymnasium—a space that, like the rest of the building, seemed to hold its own secrets. The gym was located on the lower level of the school, beneath the first floor. As we entered, Nancy pointed out a few key areas: to the right were three small rooms—a girls' bathroom, a boys' bathroom, and what used to be the coach's office. On the left side of the gym, there was a doorway that led to the boiler room, which then connected to the fuel room where the furnace's fuel was stored. The boiler room, she explained, had once been a hub of activity for the janitors, a place where they spent a significant amount of time during the school's operation.

As we stood in the gym, taking in our surroundings, I noticed a set of bleachers on the left side of the room, pressed up against the wall. They were several rows high, the kind that could easily seat dozens of students during a school event. We decided to take a few group photos there, with Nancy as our impromptu photographer. As I climbed onto the bleachers, something caught my eye—a metal bracket fastened to the wall behind us. It was an unusual sight, about three feet wide and extending roughly five to six feet high. The bottom of the bracket was positioned six feet off the ground, a detail that stuck in my mind, especially after I later had Bondsy, who stands 6'1", stand beneath it for height comparison purposes. His head barely reached the bottom of the bracket, which made me even more curious about its purpose.

Little did I know at the time that this seemingly insignificant observation would take on much greater significance as the night progressed. The gymnasium, like the rest of the school, was full of mysteries—some of which we were about to uncover.

When Nancy finished her walkthrough, we began the methodical process of setting up our equipment, while she returned to the area that she and her husband, Jim, had converted into their living quarters. The building was now ours for the night—a silent, looming structure that held secrets we were determined to uncover.

Our first task was to strategically place Sony audio recorders throughout the building, ensuring that every creak, whisper, or unexplained noise would be captured. Next, we turned our attention to setting up the infrared surveillance cameras. I had a routine for this, a procedure I had followed for years without question. The cameras were placed in key locations, their angles meticulously adjusted to cover the most active areas. Only after all the cameras were in place and the angles set did I start the recordings, ensuring that the timestamps were synchronized. This approach, while logical, would soon reveal a flaw that would make me rethink how I conducted future investigations.

We set up our command center in the second-floor hallway, a narrow, dimly lit space that offered just enough room for a folding table, a video monitor, and a DVR. The first three infrared cameras were already in place and connected to the system. One was on the first floor, pointed directly at a row of lockers, their doors ominously wide open. This was no random choice; previous investigators had reported these doors slamming shut on their own, and I wanted to capture that moment if it happened again. The second camera covered the principal's office, a room steeped in unsettling history and dark rumors. The third camera was in the gymnasium, its lens focused on the bleachers along the left wall—the very spot where I spied the mysterious metal bracket earlier in the evening.

At that moment, Bondsy was on the first floor, while Chris was in the men's bathroom adjacent to the gym. Sarah stood a few feet from me as I connected the fourth and final camera to the DVR. This camera had been placed on the third floor, its watchful eye monitoring the long, empty hallway that stretched into darkness.

As I secured the final connection, I glanced at the video monitor, checking the feeds from cameras one, two, and three. Everything seemed normal, the screens displaying the familiar static images of the deserted school. But then something on the feed from the gymnasium caught my eye—something that made me pause and take a closer look.

At first, I thought my eyes were playing tricks on me, but as I stared at the monitor, the dark shape became clearer. Moving across the center of the gym floor, directly in front of the bleachers, was a tall, black, shadowy figure. This was no ordinary shadow, no mere trick of the light or reflection. It was the unmistakable form of a person, but there was something profoundly wrong about it. The figure was impossibly tall, its limbs elongated as it swung its arms and legs with a deliberate, almost purposeful gait.

I knew immediately what I was seeing—the shadow man Nancy had described during the walkthrough. This was not just a shadow cast by an entity; this shadow *was* the entity. It moved with intent, crossing the gym as if on a mission, heading straight toward the door that led to the janitor's room. I watched, transfixed, as the figure continued its journey. Even as it reached the darker, more obscure corner of the gym, its form remained distinctly visible—darker than the shadows that surrounded it.

The camera in the gym was positioned thirty feet from the bleachers, resting on a six-inch-high tripod. With a range of sixty-five feet, the infrared camera's clearest image was in the center of the frame, where the figure was now moving. The outer edges of the scene were naturally darker, but this shadow person seemed to defy the limitations of the infrared. It was so dark, so dense, that it remained visible even in the farthest, dimmest corner of the gym.

It took several seconds for my brain to catch up with what my eyes were seeing. I quickly scanned the scene for any sign that one of my team members might have wandered into the gym without me noticing. But there was no one—no one, that is, except the shadow figure, which by now had nearly reached the corner of the gym.

I could no longer contain my excitement. "There's a shadow man walking across the gym!" I shouted, breaking the silence that had settled over the command center. Sarah, startled by my outburst, hurried over to the monitor, but by the time she arrived, the figure had vanished into the darkness, leaving only an eerie, empty gym on the screen.

What I had just witnessed was not just a shadow—it was an encounter with something that defied explanation, something that challenged every assumption I had made about the paranormal. And as the night wore on, the implications of what I had seen would weigh heavily on me, altering the course of our investigation and forcing me to confront the unknown in ways I had never anticipated.

I bolted down the stairs, my heart racing as I shouted for Bondsy and Chris. "Where are you guys?" My voice echoed through the empty corridors. I needed to know their exact locations to ensure neither of them had been in the gym when I saw the shadow figure. Almost immediately, Bondsy's voice rang out from somewhere on the first floor, confirming his location. But Chris didn't respond. I felt a pang of anxiety as I reached the gym, only to discover that Chris had been in the boy's bathroom at the time. Neither of them could have cast the shadow I had seen.

The exhilaration of witnessing the shadow man firsthand was undeniable, but it was tempered by a profound sense of disappointment. In all my years as a paranormal investigator, this moment stood out as one of the most thrilling yet frustrating experiences. I wasn't recording when I saw the shadow man—an oversight that haunted me. As I adjusted the surveillance system and fine-tuned the camera angles, I had neglected to start the recording. If I could see the figure on the monitor, it would have been captured on video, too. I had missed what could have been the evidence of a lifetime.

The shadow figure I saw bore a striking resemblance to the artist's drawing Nancy had shown us earlier, with a few notable distinctions. The artist had described the entity as extremely thin and incredibly tall, estimating it at around seven feet. As incredible as that might sound, I was confident that what I saw was at least that tall, if not taller. My certainty stemmed from the old metal bracket on the gym wall I had noticed earlier in the evening.

This bracket, made of angle iron and rectangular in shape, was likely a remnant of some long-forgotten gym equipment. It stood five feet high and three feet wide. When the shadow figure moved across the gym, passing in front of this bracket, its head appeared roughly eighteen inches above the bottom edge. Later, I used Bondsy as a human measuring stick to gauge the figure's height. Standing under the bracket, the top of Bondsy's head was just an inch above the bottom edge. This meant that the shadow man was at least eighteen inches taller than Bondsy, placing it at a towering seven to seven and a half feet tall.

The figure in the artist's drawing was thin, almost skeletal, whereas the shadow person I saw appeared more substantial. It looked as though it was draped in a long, black coat, and its head was adorned with a short-rimmed cap, reminiscent of an old sea captain's attire. But the most striking difference was its midsection. The figure seemed to have a large, protruding stomach—a potbelly that added an unsettling dimension to its appearance.

This observation raised unsettling questions: Was there more than one shadow person haunting Farrar School, or could these entities change in appearance over time, gaining or losing mass as we do? The idea that these beings could evolve, or that multiple entities could exist in the same space, added a new layer of complexity to the investigation.

About a year after this experience, in November 2014, I watched an episode of a television show called *Ghost Stalkers*. The show followed a two-man team of paranormal investigators, and in this particular episode, they were investigating Farrar School. My interest was piqued when one of the investigators claimed to have seen the shadow man in an upstairs hallway. His description caught my attention because he mentioned seeing some sort of protrusion or abnormality around the midsection. He had seen the figure moving toward him, and his account matched mine in an uncanny way. I had observed the shadow man from the side, viewing it in profile. Could this protrusion—the potbelly I had noticed—be the same feature seen from a different angle?

The encounter left me with more questions than answers, deepening the mystery of Farrar School and its enigmatic shadow people. Whether this entity was a singular presence or one of many, its existence challenged my understanding of the paranormal and reinforced the notion that we had only just begun to scratch the surface of what truly haunts these old, abandoned halls.

After the adrenaline rush from witnessing the shadow man began to fade, we regrouped and focused on completing our setup. We had a long night of investigation ahead, and the strange events were far from over. It didn't take long before we found ourselves in the midst of yet another unnerving experience.

Sarah and I were on the third floor, combing through one of the old classrooms, while Chris and Bondsy were supposed to be investigating together on the second floor—or so we thought. Unbeknownst to us, Chris had decided to head down to the gym, assuming Bondsy was following right behind him. As Chris descended the stairs, he started walking toward the gym, but after just a step or two, something made him freeze in place. Later, when we all gathered to compare notes, Chris told us he had heard something that sent a shiver down his spine—a deep, heavy exhale, like the breath of a large person standing just behind him.

Even though Sarah and I were still on the third floor, we distinctly heard Chris call out. He yelled, "Bondsy," pausing for a moment as if waiting for a response before calling out Bondsy's name again. Then, his voice took on a more urgent tone as he called out my name twice: "Larry-Larry!"

Bondsy, hearing Chris's voice echo through the building, shouted back, "What do you want?" But Chris didn't respond. Concerned, Bondsy made his way downstairs to find him. As he reached the bottom of the stairs, he encountered Chris coming back up, looking visibly shaken. Chris had decided that lingering alone in the dark with whatever had exhaled behind him was not an option. He made the quick decision to retreat upstairs, even though it meant passing by the source of the ominous sound.

By the time Sarah and I made it downstairs to see what all the commotion was about, Chris was already recounting his experience to Bondsy. He was a mix of agitation and fear—agitated because he thought Bondsy was right behind him and shaken because of the breathy sound that had stopped him in his tracks.

Chris explained how, after hearing the exhale, he turned around slowly, not knowing what might be standing there. But when he looked, no one was there. "A cold chill went down my spine, and the hair stood up on the back of my neck," he said, his voice still tinged with anxiety. "I knew what I heard, and I knew I was alone down there. Whatever it was, it was loud, and it was close."

Though he hadn't seen anything, Chris was certain that what he heard was a loud, human-sounding breath. And if the source of that breath was still there, it was between him and the staircase leading back to safety. His first instinct was to call out for us, hoping someone would respond and save him from having to walk past the unseen entity. But when no one answered, he had no choice but to muster up the courage to head back upstairs, passing through the very spot where the sound had originated.

I could see that Chris was genuinely rattled by what had happened. His encounter with whatever was at the bottom of the stairs was enough to shake even a seasoned investigator. But as unsettling as it was to hear his story, the full impact didn't hit me until I reviewed the evidence later. Personal experiences are one thing, but they take on a whole new weight when they're backed up by hard evidence—evidence that we were fortunate enough to capture.

Earlier that evening, I had placed a digital audio recorder on a table near the far side of the gym, close to the girls' bathroom. The recorder was positioned at least 100 feet from where Chris had his encounter. Several days after the investigation, as I sat down to review the audio, I was struck by the clarity of the recordings. Farrar School was one of the first locations where I used Sony recorders, and the difference in quality compared to my older RCA devices was remarkable. Even at a distance of 100 feet, the new recorders picked up sounds with incredible precision.

As I played back the audio from the moment Chris descended the stairs, the events of that night unfolded with chilling accuracy. I could clearly hear Chris's footsteps as he made his way down. Then, unmistakably, there was the sound—a loud, clear exhale, almost a sigh. The breath was so distinct that it sent a shiver down my spine, just as Chris's account had. Right after the exhale, I heard Chris stop dead in his tracks, his footsteps abruptly ceasing. The audio captured every detail, confirming everything Chris had described. The breath, the pause, the fear in his voice—it all played out exactly as he had recounted.

Listening to the recording, I could only imagine how terrifying it must have been for Chris, standing there in the dark, knowing something was behind him, yet seeing nothing. The evidence not only validated Chris's experience but also added additional intensity to the already unsettling atmosphere of Farrar School. It was one thing to hear about paranormal encounters, but when they're captured on tape, the reality of what we're dealing with becomes undeniable.

A short time later, as if the night had not already provided its fair share of eerie encounters, a truly remarkable experience unfolded in a third-floor classroom. This encounter would affirm my growing suspicion that whatever entity—or entities—haunted Farrar School was no mere aimless apparition. It was intelligent, aware, and purposeful.

The event occurred shortly after Chris and Bondsy had their own startling experience in a second-floor classroom. Their investigation had taken a dramatic turn when they were abruptly jolted by a loud, metallic banging sound that seemed to come from nowhere. Sarah and I were down the hall, engrossed in our own investigation, and we did not hear the noise. However, the startled yells from Chris and Bondsy reached us, signaling that something significant had happened.

When they rejoined us, their faces were a mixture of excitement and apprehension. They recounted their experience: they had been in a classroom at the end of the third-floor hallway, trying to locate the source of a persistent noise. One of them had remarked on the eerie silence in the room when, without warning, a deafening metallic bang erupted. It was as if someone had taken a set of metal table legs and slammed them shut with force.

The source of the noise was traced to a folded cafeteria-style table leaning against the back wall. The sound they heard was identical to the noise of metal table legs being forcefully retracted and slammed together. Bondsy, ever the meticulous investigator, had recorded the sound and played it back for Sarah and me. The recording was strikingly clear, capturing the sharp, resonant clang that had startled them. Intrigued, Bondsy and I decided to return to the classroom and see if we could coax any further activity from the elusive entity, while Sarah and Chris continued their exploration at the opposite end of the hallway.

The classroom was as quiet as a tomb when we entered. Bondsy showed me the exact spot where he and Chris had been standing when the banging noise occurred. As we reviewed the details of their encounter, an idea struck me. This was once a school, filled with children who had probably spent hours in these very classrooms. What if, by invoking the school environment, we could stir something in the spirits of those young students?

I decided to take a rather unconventional approach. With all the desks still arranged as they had been in their school days, it seemed fitting to hold a mock class. If the spirits of the children who once roamed these halls were still around, perhaps they would respond to a familiar call to order.

Bondsy, playing along, settled on the floor at the back of the room, leaning against the wall in a manner that mimicked a student's posture. I took my place at the front of the room, adopting the stance of a teacher ready to begin a lesson. Channeling my best authoritative tone, I announced, "My name is Mr. Wilson, and I will be your substitute teacher for the day. Please take your seats as it is time to start math class."

Math Classroom

The atmosphere in the room seemed to shift, a detectible tension hanging in the air as if the classroom itself were holding its breath. We waited, our senses on high alert, wondering if this playful intrusion into their former lives would elicit any kind of response from the unseen occupants of the room.

My plan was straightforward: I intended to pose simple addition and subtraction questions, and I instructed any spirits present to respond by knocking loudly enough for us to hear. What unfolded next would become one of the most extraordinary encounters in my career as a paranormal investigator.

I began with, "What is one plus one?" Within seconds, a clear *knock-knock* reverberated from the floor—a precise and accurate answer to my question. Startled, I turned to Bondsy and asked, "Did you hear that?"

"I did," Bondsy responded, puzzled. "Where did it come from?"

"It came from the floor," I replied, trying to pinpoint the sound. It was as if someone—or something—had knocked on the floor with their knuckles between where Bondsy was seated and where I stood.

The second response was equally remarkable. I posed the question, "What is one plus zero?" Instantly, we heard a single knock from the floor. The immediacy of the response left me astounded. I asked Bondsy, "Was that you?"

Without directly answering, Bondsy asked, "Did you hear that?"

"Yeah," I confirmed.

"That's not you?" Bondsy probed further.

"No," I replied firmly. "I'm standing perfectly still."

Encouraged, I addressed the unseen presence: "The first answers were correct. Can you respond again by knocking?" I then asked, "What's one plus one?" Almost immediately, and louder than before, we heard a decisive *knock-knock.*

Bondsy's excitement was apparent. "Dude, did you freaking hear that?"

"I did," I answered, equally exhilarated.

The mock classroom experiment continued for thirty minutes. Over that period, I asked twenty questions and recorded nine clear, accurate responses. Although we heard more responses than we recorded, several were too faint for the recorder to capture.

Even with the evidence mounting, it seemed almost too incredible to believe. We continually questioned each other to ensure we hadn't accidentally caused the knocking sounds. At one point, the entity—or entities—decided to drive the point home. I asked, "Last question. What is one plus one plus one?" Instead of knocks, the response came as *click, click, click*—a sound reminiscent of someone flicking their tongue against the roof of their mouth, somewhat akin to wooden blocks clacking together.

Bondsy's reaction was immediate and colorful. For the sake of keeping this account family-friendly, I'll just say his response was a stunned, "Did you freaking hear that?"

To which I asked, "Is that you?"

"No!" came the emphatic reply.

During the EVP session, I decided to delve further. I asked, "Let's try this for extra credit. I need to know if you are a boy or a girl. If you're a boy, knock once. If you're a girl, knock twice." As soon as I said "twice," a loud, single knock echoed in response, indicating that our invisible student was a boy.

Next, I inquired about the spirit's grade in school. "Can you knock and tell us what grade you're in? Just make as many knocks on the floor as the grade you're in. One for first grade, two for second." I began to say "three" for third grade, but my question was cut off when, as soon as I mentioned "two for the second grade," there were three distinct knocks in rapid succession. This unmistakably signified that our spectral student was in third grade.

"Did you hear that?" Bondsy asked, wide-eyed.

"Three—third," I clarified, referring to third grade.

"Yup," Bondsy confirmed.

What Bondsy and I experienced during this mock classroom

experiment stands as one of the most affirming interactions in my more than two-decades of paranormal investigation. The clarity and accuracy of the responses left no doubt in my mind: something intelligent was engaging with us in that classroom on October 4, 2013.

The responses were consistent with a third-grade student, typically eight or nine years old. I couldn't help but wonder if we were communicating with the same little boy seen by Nancy on the stairway. Her description of the child as approximately three and a half feet tall closely matched the height of an eight- or nine-year-old.

The manner in which the questions were answered made perfect sense. The responses to the addition questions were slightly delayed, as if the spirit were working through the calculations—a behavior fitting for a third grader. In contrast, the answers to questions about gender and grade were immediate, reflecting information the spirit would easily know.

This session was more than just a series of knocks and clicks; it was a profound confirmation of intelligent interaction with a young spirit who seemed to retain a remarkable sense of identity and knowledge from his days at Farrar School.

After our classroom experiment, we met up with Sarah and Chris and headed to the first-floor area the Olivers had set up for investigators to take breaks. We discussed our experience in the classroom and played back some of the responses we recorded.

After the break, we returned to the classroom and tried the math test again. Unfortunately, we didn't get the same results but did hear the clicking sound again. It sounded like it was coming from somewhere in the middle of our group as we sat at desks. Interestingly, it didn't start until we discussed moving to another part of the building, almost as though it was trying to entice us to stay in the room.

During the course of the night, we investigated the old janitor's room, located on the same level as the gym. Of all the places in the building, this was the only location that gave me a creepy, uneasy feeling. After all, it was in a damp and dingy part of the building, with an old boiler, a workshop area, and a reputation as a place where a janitor, alleged to have molested children, hung out.

We had been in the janitor's area for twenty minutes when we heard tapping sounds coming from above us. There was a bit of a breeze outside due to a thunderstorm in the area, so I figured the noise was caused by the old ductwork above us, vibrating and rattling. As we were discussing the tapping sound, we heard a loud noise that originated from the first floor. It sounded as if someone was upstairs moving around. The noise was loud enough to cause us to scurry out of the janitor's area and run upstairs to check it out. When we reached the first floor, the noise stopped, and we found nothing.

The last area of the building we investigated was the third-floor auditorium. We spread out, with each of us sitting on a different side of the room. At one point, Sarah sat at a table on the auditorium stage. Speaking aloud, she invited any spirits present to sit with her at the table and make noise so we would know they were around. Unfortunately, no specters took her up on her offer. We conducted a final EVP session, but neither heard nor recorded any responses. It was approaching 4:00 a.m., so we decided to call it a night and began to break down the equipment and pack up, as we had a five-hour drive ahead of us. But what a night it was.

Investigation Summary

If someone asked me to recommend a place where they would have a better-than-average chance of experiencing paranormal activity, the Farrar School in Iowa would be one of the top ten places I would suggest.

After the eerie events of October 4, 2013, there is little doubt in my mind that something profoundly unusual persists within the walls of the Farrar School. The lingering presence of former students and staff seems undeniable, suggesting that this once-vibrant institution is now a stage for those who cannot or will not move on. Perhaps these spirits cling to a time that was once full of significance and joy, or maybe something more malevolent has tainted their memories.

Could it be that the children who remain are striving to reclaim the innocence that was stolen from them? If so, I can only hope that our interaction in the classroom—a fleeting yet poignant connection—offers some measure of solace or restoration for what was lost. If these spirits are indeed remnants of those who once trusted and looked up to their caregivers, then perhaps our presence has, in some small way, helped to mend the broken threads of their past.

But what of the other figures from the school's history? If children haunt these halls, are the former teachers and staff also bound to this place? If the unsettling rumors of abuse and cover-ups hold any truth, might those involved now walk these corridors in a perpetual state of shame or remorse? Alternatively, could there be individuals whose reputations were unfairly tarnished, their names sullied by misunderstandings or wrongful accusations? If so, is it possible that their spirits linger here, seeking redemption or a chance to reclaim their tarnished honor?

The ghosts of Farrar School may not only be those of the young and vulnerable but also of those whose lives and legacies were overshadowed by darkness. In this realm where the past and present converge, the echoes of history and the cries for justice may continue to reverberate through the school's haunted halls.

Our team had a front-row seat to the peculiarities that define the Farrar School during our investigation. When Nancy first presented the drawing of the shadow man—a chilling depiction sketched by an eyewitness—I approached it with a degree of skepticism. I was cautious, holding onto the notion that eyewitness accounts are often tainted by imagination or misperception. However, my doubts would soon be shattered.

Not long after, I found myself face-to-face with what could only be described as the very entity depicted in the drawing. As it moved across the gymnasium floor, it was clear this was no mere trick of light or shadow. The figure traversed the space with an unmistakable sense of purpose, navigating around the bleachers as if it had a destination in mind. This was no ethereal wisp but a defined presence with intent.

The breath Chris encountered adds another layer to this enigma. Captured by our audio recorder from a distance of one hundred feet, the exhale was powerful and distinct—an auditory anomaly that could not be easily dismissed. The nature of the sound suggested something substantial and deliberate was behind it. Was this breath a manifestation of the shadow man, or did it belong to a different entity entirely? Was the intent behind it malicious, or was it a form of ghostly mischief aimed at unsettling Chris? Regardless of its purpose, the sound unmistakably indicated that whatever was responsible knew of Chris's presence.

The impromptu classroom experiment with Bondsy remains one of the most profound encounters in my more than two decades of paranormal investigation. The clarity of the interaction, coupled with the precise responses we recorded, confirmed that we were not alone in that classroom. The experience was both validating and electrifying, an affirmation of the intelligent presence we had suspected all along.

Investigating the supernatural is as enlightening as it is humbling. Each encounter offers a fleeting glimpse into a dimension that, for most, exists only in movies and make-believe. These moments challenge our understanding of reality and confront the boundaries of our accepted beliefs.

People often ask why I continue to delve into the paranormal after so many years. In the early days, it was undoubtedly the curiosity and the thrill of the unknown. But now, having witnessed the profound and bizarre events at places like Farrar School, the attraction has changed. It's no longer just about the adventure; it's about the inexhaustible quest to understand a reality that defies explanation. The more I uncover, the more I realize that the strangeness of the paranormal is endless, and Farrar School stands as a testament to just how extraordinary it can be.

MALVERN MANOR
MALVERN, IOWA

FIVE

On the night of August 9, 2019, I led Bondsy and two of his interns, Lauren and Kate, to one of Iowa's most notoriously haunted locations—Malvern Manor. For me, taking Bondsy and his team on these investigations offers more than just camaraderie; it's an opportunity to provide firsthand experience of the unexplainable to those who remain skeptical about the existence of the paranormal. After two decades of exploring the unknown, I've witnessed phenomena that defy logic, and sharing these moments with others can be profoundly eye-opening.

This investigation carried added significance—it was the first time that interns Lauren and Kate were stepping into the shadowy world of paranormal exploration. In my experience, it's often the newcomers who find themselves at the center of the most intense interactions. There's something about fresh energy, unjaded by countless nights of ghost hunting, that seems to draw the attention of spirits.

As we approached Malvern Manor, it was clear that the interns viewed the night as more of an adventure than a serious investigation. The prospect of spending the night in a creepy old building held a certain thrill, but they hadn't yet grasped the gravity of what we were about to undertake. That's the thing about places like Malvern Manor—they lull you into a false sense of security, only to reveal their true nature when you're least prepared.

What unfolded that night in 2019 would serve as a lesson they wouldn't soon forget. The two interns, who had started the evening with casual curiosity, would leave Malvern Manor with a newfound respect for the supernatural. The events that were about to transpire would strip away any remaining doubts about the reality of the paranormal. By the time the night was over, Lauren and Kate would understand that the forces at work in these haunted locations are not to be underestimated.

History and Haunting

Malvern Manor, with its weathered façade and quiet corridors, carries the weight of history within its walls. Built in the late 1800s, it began as a bustling family-run hotel, catering to the steady stream of traveling salesmen who relied on the nearby railroad for their journeys. In its heyday, the manor was a beacon of hospitality, offering weary travelers a place to rest before they continued on their way. But time has a way of changing the landscape, and with the rise of the automobile, the reliance on rail travel waned. The hotel, once a thriving business, eventually closed its doors, leaving the grand building to stand in silence, its halls empty of the life that once filled them.

By the mid-1900s, Malvern Manor had transformed from a hotel into the personal residence of the Gibson family. T.D. Gibson and his wife took on the responsibility of raising their nieces and nephews after the children's biological parents were deemed unfit to care for them. The arrangement, while intended to provide stability, introduced its own set of emotional challenges, especially for the youngest child, Inez.

Inez was particularly affected by the separation from her parents. The move to her aunt and uncle's home, meant to be a refuge, instead became a place of quiet despair for the young girl. The weight of her emotions, coupled with the pressures of adjusting to her new life, began to take its toll.

The tragic story of Inez's death has become one of the most unsettling legends associated with Malvern Manor. According to accounts, one day Inez told her brother Otto that she was going outside to jump rope. But when Otto later entered her room, he was confronted with a scene that would haunt him forever—his sister hanging from the closet, the jump rope cruelly wrapped around her neck. The official ruling was that her death was accidental, but rumors quickly spread through the small town, suggesting that the true cause was far more insidious—emotional stress brought on by the trauma of her separation and a perceived failure at school.

The tragedy of Inez has left a lingering mark on Malvern Manor. Visitors to the home have reported hearing the faint, disembodied voice of a child echoing from her room, a sorrowful reminder of the young life that was cut short. Some believe that Inez's spirit remains trapped within the walls of the manor, forever tethered to the place where her innocence was lost. Whether it's the restless soul of a child seeking comfort or something more unfathomable, the story of Inez is a chilling testament to the dark history that continues to haunt Malvern Manor.

After the Gibson family vacated Malvern Manor, the building underwent yet another transformation, this time into a combination convalescent home and minimum-care facility. What was once a place of hospitality and later a family residence now became a repository for those society had largely forgotten. The manor's new residents were a diverse group, ranging from alcoholics battling their inner demons to schizophrenics lost in the labyrinth of their own minds. It was a facility meant to provide care, but the reality was far grimmer.

The very nature of the manor's patients—each grappling with their own set of complex, often misunderstood disorders—meant that the care they received was inconsistent at best. Stories began to circulate about the conditions within the facility. Some whispered that the caretakers were overwhelmed, unable to give the patients the attention and support they desperately needed. Others suggested that the manor was a place where the mentally ill and the addicted were simply warehoused, left to languish in their own private hells.

This neglect, whether through incompetence, apathy, or a simple lack of resources, left a deep scar on the manor. Many believe that the suffering experienced within those walls has lingered long after the facility closed its doors. The manor's past, steeped in pain and neglect, seems to have given rise to something far more sinister—a haunting that refuses to be silenced.

Visitors to Malvern Manor have reported unsettling experiences that defy logical explanation. Some have seen shadowy figures lurking in the corners, while others have felt an overwhelming sense of despair that seems to seep from the very walls. The air inside the manor is thick with a presence that can only be described as perceivable—a lingering echo of the lives that were once confined there.

The idea that the manor's dark history has contributed to its current state is not far-fetched. In the world of the paranormal, it is often believed that intense emotional suffering can leave an indelible mark on a place, creating a kind of psychic residue that attracts and traps spirits. If this is true, then Malvern Manor is a prime candidate for such activity. The alleged lack of care, the suffering, and the despair of its former residents may have woven together to create a tapestry of haunting that still plays out in the manor's shadowy halls.

The Investigation

As we embarked on the long, six-hour, four-hundred-mile journey to Malvern Manor, Bondsy, always eager for a good story, asked me to recount some of the most inexplicable things I'd witnessed over the years. The miles seemed to pass quicker as I dove into a tale about another notorious Iowa haunt, the Villisca Axe Murder House, located just thirty-eight miles from our destination. I recounted how, after returning home from that investigation, I began hearing a strange voice whisper my name in my right ear. It was an experience that defied explanation and left a permanent mark on my psyche. Little did I know that this particular story would soon take on new significance.

We were scheduled to meet the owner of Malvern Manor, Kurt Fricke, at 6:00 p.m., but we arrived in town early. With some time to kill, we decided to grab supper at a local diner. As we waited for our food, the conversation took an unexpected turn. Kate, one of the interns, leaned over to Lauren and said, "Did you tell Larry what happened to you?"

"No, I almost forgot," Lauren replied, her tone a mix of excitement and unease.

Intrigued, I asked Lauren what had occurred. She hesitated for a moment before speaking.

"Well," she began, "it was really weird. Do you remember when you were telling that story about the murder house and how you heard your name whispered in your ear?"

"Of course," I said, wondering where this was headed. "Why?"

Lauren's eyes widened as she continued. "Well, after you finished the story, I shut my eyes to take a nap. I wasn't fully asleep when I heard a voice whisper my name in my ear. It was freaky!"

I leaned in, feeling a familiar chill run down my spine. "Which ear did you hear the whisper in?" I asked, already suspecting the answer.

"My right ear," she replied, her voice trembling slightly.

The room seemed to grow quieter as I absorbed her words. What Lauren experienced wasn't new to me—it had happened to others who had listened to my stories about the Villisca house. Each time, the whisper was heard by the right ear, the same ear I would hear my name called. This eerie connection was hard to dismiss as mere coincidence.

I believed Lauren, not just because her story aligned with my own experiences, but because I had seen this pattern emerge before. Hearing that whisper, however, was only the beginning of the strange events that would unfold for Lauren on this fateful night.

When supper was over, it was time to meet the man who held the keys to Malvern Manor's mysteries. We made our way to the manor, where Kurt greeted us with a firm handshake and a welcoming smile. Without wasting any time, he began to lead us through the building, offering a brief yet unsettling tour. Our first stop was the nurses' wing, a section of the manor that had seen its fair share of suffering and, if the stories were to be believed, something far darker.

At the end of the hallway leading to the nurses' wing, we arrived at an old nurses' station. The counter, once bustling with activity, now stood silent, its storage bins still holding the echoes of the medical charts that were once meticulously filed there. The atmosphere was thick with the weight of history, the kind that clings to your skin and leaves you with an uneasy feeling.

As we reached the convergence of the main hall and the nurses' station, Kurt directed our attention to another hallway, known ominously as the Shadow Man Hallway. The name itself sent a chill down my spine. According to Kurt, this was the domain of Malvern

Manor's infamous shadowy phantom, a figure often seen lurking in the dark corridors of this part of the building. Those who had encountered the shadow man spoke of a malevolent energy, a presence that could turn even the most seasoned investigator's blood cold.

Kurt shared the unsettling experiences of several female investigators who claimed that a shadowy figure had followed them out of the hallway and into the main part of the manor. As he spoke, I couldn't shake the feeling of being watched, a sensation that crept over me with every step we took deeper into the nurses' wing. There was something about this place, especially the Shadow Man Hallway, that made the hair on the back of my neck stand on end. It was as if the very walls were alive, observing our every move.

The strangeness wasted no time in making itself known. Bondsy, always diligent, was recording video on his cellphone as Kurt described the activity that had been reported in the hallway. Suddenly, Bondsy interrupted Kurt mid-sentence, his voice tinged with a mix of surprise and curiosity. He had caught something unusual on his phone—a bright white light that had moved in front of his camera seemingly out of nowhere.

We crowded around the small screen as Bondsy replayed the video. Sure enough, there it was—a white light, vivid and inexplicable, drifting across the frame. What struck us as particularly odd was the fact that the doors to the adjacent rooms were closed, plunging the hallway into darkness. There was no apparent source for the light, no logical explanation that could account for its sudden appearance.

Having reviewed hundreds of hours of video from various investigations, I'm well acquainted with the usual suspects—dust, moisture, pollen particles—those mundane anomalies that can trick the eye and the camera. But this was different. The light that Bondsy recorded wasn't a trick of the environment; it was something else entirely. It was as if a piece of the unknown had momentarily revealed itself, then vanished just as quickly.

After the brief but intense excitement, we continued our journey down the hallway, the air thick with anticipation. Our destination was Room Seven.

To reach Room Seven, you have to make a right at the nurses' station and walk down the dimly lit corridor until you come to the last room on the right-hand side. Before we stepped inside, Kurt paused, his expression turning somber as he prepared to share one of Malvern Manor's more tragic tales—a story that still resonates in the very walls of this room.

He spoke of a woman, once exquisite in both appearance and spirit, who found herself committed to the manor when it served as a home for the mentally ill. Her presence, many believe, still lingers in this room, trapped by the torment that led to her untimely confinement.

The woman's descent into madness began at the behest of her husband, who sought help after she began causing physical harm to herself. By all accounts, she had been healthy and happy, a devoted wife and mother with a life that seemed picture-perfect. She was known for her long, beautiful hair, a symbol of her elegance and grace. But something shifted inside her—a peculiar , deep-seated belief that she was no longer attractive, and her husband's love had faded. This obsession with her appearance consumed her, and she would spend hours in front of a mirror, brushing and pulling at her hair until it fell out in clumps.

Despite her husband's efforts to reach her, to reassure her of his love, the woman spiraled deeper into despair. Her fixation on her appearance grew so intense that her husband, seeing no other option, made the heart-wrenching decision to have her committed to Malvern Manor.

During her time at the manor, the staff often found her standing in front of the mirror in her room, endlessly combing and tearing at what remained of her once-beautiful hair. The mirror became both her tormentor and her only confidant, reflecting back the image of a woman she no longer recognized.

Even now, long after her passing, the room remains charged with her tragic energy. Paranormal enthusiasts who have ventured into Room Seven have reported capturing the sound of a woman's voice on their recordings—a mournful echo from the past. Others have witnessed the door opening and closing on its own, as if the spirit of the woman is still trapped in the cycle of her despair.

Later, you will read about two incidents that happened during our investigation related to intern Lauren. The incidents took place at the opposite end of the hallway and may be connected to the spirit of the woman in Room Seven.

As our tour of the Shadow Man Hallway concluded, we made our way to the second floor to continue our walkthrough of Malvern Manor. When it comes to investigating locations that charge a fee, I approach them with a healthy dose of skepticism. Too often, the lure of profit can overshadow the genuine pursuit of the unknown. But it was during this part of our walkthrough that the manor began to challenge my doubts and make me reconsider.

We climbed the creaking wooden stairs to the second floor, where Kurt paused, turning to face our group with a serious expression. There was an intensity in his eyes that hadn't been there before, a kind of anticipation that piqued my curiosity.

"Before we go any further," Kurt announced, his voice echoing slightly in the narrow hallway, "I want to disclose something important. I guarantee that something will happen to at least one of you when we enter a particular room on this floor. To prove it, I'll take one of you aside right now and tell you exactly which room it will be and what will happen—without revealing it to the others."

I couldn't help but raise an eyebrow at his claim. It sounded like the kind of showmanship I've encountered at other supposed haunted locations, where the goal is to ramp up the tension and sell the experience. "This should be good," I thought to myself, skepticism creeping in. I've seen too many places where the promise of a paranormal encounter was nothing more than a well-rehearsed parlor trick.

But as I stood there, waiting to see how the scene would play out, there was a part of me that wondered if Malvern Manor might just be different. The air on the second floor felt thick with something I couldn't quite place—something that made me uneasy in a way that wasn't easily dismissed. I had the sense that whatever Kurt was about to reveal might push this investigation into territory I hadn't anticipated.

Kurt chose Bondsy as the one to confide in, pulling him aside for a brief, private conversation before we continued the tour of the second floor. As we resumed, the atmosphere felt charged with anticipation, each of us quietly wondering what Bondsy had just learned.

At the top of the stairs, the first room on the right is a small, unassuming space known as Hank's Room. Kurt explained that this was once the quarters of a particularly troublesome patient named Hank, a man notorious for his violent tendencies toward women, especially the nursing staff. The room itself was nondescript, its bland walls contradicting the sinister history Kurt recounted.

I didn't sense anything unusual at first. The room seemed to be just another relic of the manor's troubled past. But without warning, Bondsy turned to intern Lauren and asked, "Why are you holding your stomach?"

Lauren looked up, a slight frown creasing her forehead. "Because I have pains in my stomach," she replied, almost as if she hadn't realized it until that moment.

"How long have you had them?" Bondsy pressed.

"They just started," Lauren said, her voice tinged with surprise.

"My stomach hurts too," intern Kate chimed in. "But I think it's because of the ranch dressing I had on my dinner salad—it didn't agree with me."

Kurt continued his narrative, detailing more of Hank's unsettling behavior, but the girls' discomfort was evident. As we exited the room, Bondsy asked us all to stop in the hallway. There was a seriousness in his tone that commanded our attention.

"Before we go any further," he began, "I need to tell you something. When Kurt pulled me aside earlier, he told me that one or more of us would start feeling a stomachache when we entered Hank's Room."

The words hung in the air, and for a moment, the only sound was the creaking of the old manor settling around us. Lauren's eyes widened in disbelief. "No way," she whispered.

Kate's expression mirrored the shock. "Oh my," she muttered, her hand still pressed against her abdomen.

We all exchanged glances, the weight of what had just happened sinking in. None of us, except for Bondsy, had known what Kurt had predicted. And yet, the girls' sudden, unexplained pain was undeniable. It wasn't the result of any chemicals or strange smells—nothing in that room had physically changed. If it had been something environmental, it would have affected Bondsy and me as well, but we felt nothing.

In that moment, my skepticism began to waver. The stories Kurt had been telling us started to take on a new level of credibility. The manor, with its long history of suffering and violence, seemed to be asserting its presence, making itself known in a way that was impossible to ignore.

The next stop on our tour was a room that would soon prove pivotal to our investigation—Grace's Room. Of all the spirits said to haunt Malvern Manor, none is more infamous than Grace. Her story is one of tragedy and complexity, a narrative that leaves an indelible mark on anyone who dares to explore the space she once occupied.

Grace was not just another patient at the manor; she was a permanent resident, having spent much of her life confined within its walls. Diagnosed with schizophrenia and plagued by multiple personality disorder, Grace's mind was a perplexity of identities—some accounts claim she harbored as many as fifty-six distinct personalities. The staff who cared for her often-recounted chilling experiences, particularly the unsettling moments when a deep, gruff voice would emanate from her room, chanting ominously, "The Devil is coming to get me."

What made these episodes even more disturbing was that when the staff rushed into her room, expecting to find a man, they instead found Grace, her fragile body sitting in the corner, eyes vacant yet alive with the personality that had temporarily taken control. The voice they heard—the one that sounded so convincingly like a man—was hers, one of the many personas that haunted her fractured mind.

Grace's room remains largely unchanged, as if she might return at any moment. Her old sunglasses rest on a bedside table, a tangible connection to the woman who once wore them. Her wheelchair sits abandoned, a silent witness to the years she spent confined within these four walls. The room has an eerie stillness, a sense that time has frozen, waiting for Grace to reclaim her place.

Later in this chapter, you'll read about the events that transpired in this very room—experiences that would leave intern Lauren questioning the boundary between the living and the dead. But as we stood there during the walkthrough, none of us could have predicted the impact this room would soon have, especially on Lauren.

Kurt then led us to a room that carries with it a somber and unsettling history—the room where young Inez Gibson was found hanging in her closet. The air seemed heavier as we entered, weighed down by the tragic energy that still lingers. It was in this very space that Kurt shared a story that both intrigued and unnerved us—a story about a doll known as "Number One," which some believe is possessed.

The tale of Number One began with a paranormal team who had visited Malvern Manor with the intention of contacting Inez's spirit. As a trigger object, they placed the doll on a dresser in her room, hoping to provoke a response. What happened next defied explanation: according to the team, the doll inexplicably flew off the dresser and landed in the center of the room. But that wasn't all. When they reviewed the audio from their investigation, they were stunned to hear a voice—one they believed belonged to the doll itself—uttering the words, "Number One."

The legend of Number One didn't end there. Kurt explained that more recently, people who have dared to touch the doll have reported experiencing a string of misfortunes. Though he didn't delve into specifics, his ominous tone suggested that these were not trivial incidents. The doll, now housed in a glass case in the manors office, had become something of a cursed object.

Curious as always, I wanted to see the doll for myself. Kurt led us to the office, where the doll was displayed like an artifact, encased in glass for safekeeping. He refused to touch it himself, but with a sly grin, he offered us the chance to hold it if we dared. Bondsy, Lauren, and Kate immediately declined, their unease apparent. But I, ever the skeptic from the "I have to see it to believe it" school of thought, couldn't resist. "Sure, I'll hold it," I said, more confidently than I felt.

As I held Number One, I reminded myself that moments before Kurt held another supposedly possessed doll he purchased from Costa Rica without any ill effects. What harm could come from holding this one? Or so I thought. It wasn't long before I began to question my decision. Less than twenty minutes after holding the doll, I took a nasty fall from the top of the attic stairs to the bottom, landing hard and earning myself a sizable bruise on my hip.

The skeptic in me wanted to dismiss the incident as nothing more than clumsiness, but the timing was hard to ignore. Did Number One play a role in my tumble down the steps, or was it merely a coincidence? I'm still not sure. But as the night unfolded, a series of strange occurrences would make me reconsider the possibility that the doll was more than just a creepy artifact. The jury may still be out on Number One, but the events that followed left little doubt that something was at play—something I would come to grapple with as the investigation continued. More on this as the chapter unfolds.

The Investigation

With Kurt's departure, the manor was ours for the night, and the air seemed to crackle with anticipation. Our first order of business was to set up the audio and video equipment throughout the building, ensuring we had comprehensive coverage of every possible angle. Once our gear was in place, we made our way to the nurses' wing and the notorious shadow man hallway to kick off our investigation.

Kurt's chilling account of a shadowy figure pursuing female investigators, combined with online reports detailing the hallway as a hotspot for malevolent activity, made this area our initial target. It was clear that if we were to encounter anything truly unsettling, it would likely manifest here.

Given that this was the interns' first foray into paranormal investigation, we decided to stick together for the early part of the night. We traversed the dimly lit first-floor corridor, heading toward the nurses' station before turning left. The hallway, once bustling with the lives of patients, now stood eerily quiet. At its far end was a locked exit door, an old bedpan, and a cushioned ottoman. The tile floor in this area was pristine, devoid of debris—an important detail that would become significant sooner than later.

We didn't have a meticulously detailed plan for investigating the shadow man hallway, but after a brief strategy session between

Bondsy and me, we formulated a tentative approach. Given Kurt's account of the shadow man's interactions with female investigators, it seemed logical that our interns—Lauren and Kate—might be the focus if the entity were to manifest.

Our plan was simple yet daring. We would station the interns at the end of the hallway, using them as enticement to attract any lurking presence, while Bondsy and I maintained surveillance from the opposite end. Positioned a hundred feet away, we kept the girls within the range of our flashlights, allowing us to respond quickly if needed.

I brushed off the ottoman, removing any dust, and Lauren and Kate took their places, with Kate sitting to Lauren's right. For reassurance, Kate clung to Lauren's right arm, while Lauren held a flashlight. After ensuring the interns were comfortable with our setup, Bondsy and I retreated to our vantage point.

From our position, I shone my flashlight toward the girls, illuminating them in the dim corridor. The plan was set into motion, and it wasn't long before the atmosphere began to shift.

After only a few minutes, we heard the girls whispering among themselves when Lauren's voice suddenly cut through the quiet, tinged with anxiety. "Hey, Larry!" she called out.

"Is something wrong?" I asked, sensing the unease in her tone.

"Maybe," she replied. "Something is playing with my hair!"

"What?" Bondsy interjected, his skepticism wavering.

"Something just pulled my hair!" Lauren's voice was edged with panic.

"No way!" Bondsy exclaimed, his disbelief evident.

"Larry, can you come down here?" Lauren's plea was urgent, her nervousness palpable.

"Sure," I responded, exchanging a concerned glance with Bondsy. We hurried down the hallway, our footsteps echoing in the tense silence.

As we reached the interns, we were met with a startling scene. Kate was visibly shaken, caught between tears and laughter, her emotions in turmoil. I turned to Lauren, whose long hair had been neatly tied back in a ponytail. She explained that she had felt someone tugging at her hair, a sensation that started gently and then escalated to a sharp yank.

I directed my flashlight toward Lauren and immediately noticed a strand of her hair sticking out stiffly on the left side of her head—the opposite side from where Kate had been sitting. It was as if an unseen force had deliberately pulled her hair.

Lauren recounted feeling a distinct touch on her hair just before it was yanked. The incident left her visibly shaken, and the atmosphere in the hallway felt charged with detectable energy. The unsettling experience had clearly rattled both interns, and as we regrouped, the sense of unease lingered, underscoring the reality of the forces at play in Malvern Manor.

When the initial shock of Lauren's experience had subsided and the girls had composed themselves, Bondsy volunteered to take a seat on the ottoman while the rest of us retreated to the far end of the hallway. The plan was simple: we would observe from a distance, hoping that Bondsy would draw any lingering presence out.

As we settled into our position, I noticed a small yellow plastic ball—about the size of a baseball—lying innocently on the floor. To prevent any accidental kicks or disturbances, I carefully placed the ball in the corner behind us, out of the way. The significance of this small act would soon become clear.

The corridor remained eerily quiet as we stood in tense anticipation. Bondsy, sitting alone on the ottoman, suddenly broke the silence. "Did you guys hear that?" he asked, his voice tinged with confusion.

"No, what did you hear?" I inquired, straining to catch any sounds amidst the stillness.

"It sounded like a ball bouncing!" Bondsy replied, his tone hinting at unease.

"No, I didn't hear anything," I responded, my flashlight cutting through the darkness as I made my way toward Bondsy, scanning the hallway.

As I passed the nurses' station and directed my flashlight down the corridor, the beam illuminated an unexpected sight. About ten feet in front of Bondsy, on the otherwise empty floor, was what appeared to be a ball. "Is that a ball in the middle of the floor?" I asked, pointing my light at the object.

"No way," Bondsy muttered in disbelief. The silence was broken by an exclamation from one of the girls, still visibly shaken from the earlier incident. "Oh my," she gasped, her voice trembling.

"It is a ball," Kate confirmed, her voice a mix of shock and curiosity.

As I moved closer, the ball revealed itself to be a white ping-pong ball. The perplexing thing was that there had been nothing on the floor when we walked from the ottoman just moments before. Yet, here it was, perfectly positioned in the middle of the hallway.

"That wasn't there before," Kate asserted, her eyes wide with astonishment.

"No, it wasn't," Lauren agreed, echoing Kate's disbelief.

We gathered around the ball, trying to piece together the mystery of its sudden appearance. I turned to Bondsy and asked about the sound he had heard. "It sounded like a ball bouncing, and now there's a ball right in the middle of the floor that definitely wasn't there before," he said, his voice rising with excitement and a hint of agitation.

I picked up the ball and examined it closely. Its flat bottom suggested it had been stepped on or compressed in some way, indicating that it couldn't have simply rolled or bounced into the center of the hallway on its own.

The appearance of the ball added another layer of intrigue to our investigation, raising more questions than answers.

"All right," I announced, "it's my turn to take a seat on the ottoman. You guys head to the other end of the hall."

Bondsy and the girls made their way to the far end of the corridor, leaving me alone in the dimly lit expanse. I flicked on my flashlight to locate the ottoman and, to my surprise, spotted a shiny penny resting on the right side. This was curious because I distinctly remembered brushing off the ottoman earlier, and a penny had certainly not been there.

"Hey, does anyone remember seeing a shiny penny on the ottoman?" I called out.

"What?" Bondsy responded, his voice tinged with disbelief. "There was no penny there."

The trio, their curiosity piqued, quickly walked back toward me, their faces illuminated by the beam of my flashlight, which now highlighted the penny in question.

"That definitely wasn't there before," Lauren said, her eyes wide with surprise.

"No way," Bondsy echoed.

"Maybe it fell out of your pocket, Bondsy," Kate suggested, trying to make sense of the situation.

"It couldn't have," Bondsy countered. "I don't carry change. Remember, I told the waitress at the restaurant to keep the change. I don't like having coins in my pocket."

We all agreed that the penny had not been on the ottoman earlier, adding yet another tier to the night's eerie occurrences. Bondsy and the girls returned to their spot at the end of the hall, and I took my place on the ottoman, bracing myself for whatever might come next.

Only a couple of minutes had passed when I heard the excited chatter of Bondsy and the girls. Their voices were tinged with a mixture of excitement and apprehension.

"Did you kick that ball, Bondsy?" Kate's voice carried a nervous laugh.

"No, I didn't," Bondsy replied.

"Oh no, here we go again," Kate exclaimed, her voice trembling slightly.

"What's going on?" I shouted down the hallway, concern rising in my tone.

"See that yellow ball over there on the other side of the hall?" Bondsy's flashlight beam danced on the object. "It just rolled out from behind us on its own and rolled down the hall."

I quickly moved toward the source of the commotion. As Bondsy's light illuminated the ball, I recognized it as the same yellow ball I had placed in the corner just minutes earlier. I joined the group, who were now gathered around the ball, and asked them to show me where they had been standing before it rolled.

They positioned themselves about three feet in front of the corner where I had left the ball. Given that the floor was perfectly level, it was impossible for the ball to have moved on its own. For the ball to have rolled that distance, Bondsy or one of the girls would have had to have kicked it backward off the wall behind them with considerable force.

As we scrutinized the scene, it became increasingly clear that unseen forces were at play in the shadow man hallway. Objects were moving, and, even stranger, hair was being tugged—all pointing to a presence or phenomenon we couldn't explain. The night was unfolding with a series of unsettling events that defied rational explanation, and the shadow man hallway was proving to be the epicenter of this disturbing activity.

Next, we headed to the second floor to continue our investigation. The corridor was still, cloaked in an eerie silence that seemed almost too serene. Kate and I moved ten feet ahead of Lauren and Bondsy, our footsteps echoing softly against the old floor. Then, the silence was shattered by a loud, jarring noise—a sound unmistakably like a door being forcefully kicked.

I turned to Kate. "Did you hear that?"

"Yeah," she replied, her face reflecting a mix of curiosity and concern.

Kate and I exchanged uneasy glances, but Bondsy and Lauren remained oblivious, their attention focused elsewhere.

We immediately began to investigate the nearby rooms, scrutinizing every nook and cranny for any signs of disturbance or an explanation for the noise. Despite our thorough search, we found nothing that could have caused the bang.

It was only later, during the audio review, that the full extent of what had transpired became evident. The recording from 9:50 p.m. revealed a sequence of two distinct EVPs that captured the eerie aftermath of the commotion. When the EVPs were recorded, we heard nothing. In the first clip, I can be heard asking Kate if she heard the noise, to which she responds with a clear, "Yeah!"

But then, as if in response to our questioning, a male voice whispers, "No!" Immediately afterward, another voice chimes in with a taunting, "Well, Kate, you're not going to prove it!" The voice seemed to challenge Kate, suggesting she was too afraid to investigate further.

It was striking enough that the disembodied voice knew Kate's name—a personal detail that seemed to affirm an unsettling level of awareness. The voices were not just random echoes; they seemed to address us directly, with a familiarity that was both unnerving and intriguing.

Over the years, I've encountered this phenomenon repeatedly: names being spoken, sometimes even when I was alone. It raises an unsettling question—how could these entities, whatever they are, know our names? Whether I was alone in a room or in the company of others, there were moments when names were spoken with a clarity that defied logical explanation.

The nature of these encounters remains uncertain, but the recurring theme is an eerie intelligence behind the voices, an awareness that transcends ordinary experience. Who or what are these entities that our equipment records, and how do they acquire such knowledge in the absence of an apparent physical presence?

After our exploration of the second floor, we descended back to the main level. Kate needed to make a phone call to her college roommate in preparation for the upcoming semester. Given the lack of cellphone signal inside the manor, she stepped outside to make the call, with Bondsy accompanying her for added security. This left Lauren and me to continue our investigation alone.

We returned to the Shadow Man hallway, armed with a video camera equipped with infrared capabilities. I set up the camera near Room 7, positioning it to capture the ottoman at the end of the hall where Lauren had previously experienced her unsettling encounter. As we prepared for the investigation, I mentioned my regret at not having a camera focused on Lauren during the hair-pulling incident. To my surprise, Lauren volunteered to sit alone on the ottoman while I monitored the camera—a bold move for someone new to paranormal investigations, especially after her recent experience.

I escorted Lauren down the hall and watched as she settled onto the ottoman. I then made my way to the opposite end of the corridor to monitor the live feed from the camera. Through the viewfinder, I noticed a shadow moving behind Lauren. It was difficult to determine if it was merely her own shadow or something more sinister. I asked her to move her head back and forth, then used the telephoto lens to zoom in. It became clear that the shadow was indeed cast by Lauren herself.

As I zoomed in, I observed Lauren fidgeting and turning on her flashlight to look behind her. Before I could react, she called out, "Hey, Larry, something is touching my hair."

"You're kidding," I replied, trying to grasp the situation.

"No, it just happened again. Now something is pulling on my ponytail," Lauren insisted.

Through the viewfinder, I saw Lauren's head jerk violently backward. "Larry, can you come down here? Something just yanked on my ponytail!" she shouted, her voice a mix of urgency and calm.

"I'm on my way," I responded, quickening my pace toward her.

When I reached Lauren, she appeared remarkably composed despite the disturbance. I asked her to describe the sensation. She likened it to the way her sister used to tug at her hair when they were children. "I felt something touching my hair before it yanked on my ponytail," she explained.

Moments later, Bondsy and Kate returned from Kate making her phone call. We recounted the incident to them then headed outside to cool off and let Lauren reflect on what had happened. Once we regrouped, we proceeded to Grace's room to resume our investigation, the evening's unsettling events adding tension to our continued exploration of the dark manor.

According to Kurt Fricke, Grace's room was reputed to be among the most paranormally active spaces in the entire manor. Given the intense activity we had just experienced in the Shadow Man hallway, our expectations were heightened and tinged with a mix of anticipation and trepidation.

As we gathered in the room, Bondsy decided to take an unconventional approach. He chose to sit in Grace's wheelchair, perhaps hoping it might provoke a response from the room's resident spirit. It didn't take long before Bondsy was visibly unsettled, expressing a profound sense of discomfort. He couldn't quite articulate why, but he felt a strong urge to vacate the chair.

"Do you think you're disrespecting Grace by sitting there?" I asked, trying to understand the source of his unease.

"Yes, exactly!" Bondsy replied, his voice tinged with discomfort and conviction.

Intrigued by his reaction, I thought, "Why not?" and decided to take a seat in the wheelchair myself. Almost immediately, I experienced the same disquieting sensation—a feeling that I was somehow disrespecting Grace. Reluctantly, I, too, vacated the chair.

As I moved away, Lauren suddenly complained of feeling uncomfortably hot. This struck me as odd, considering the air conditioning was running and the room was relatively cool compared to other parts of the manor. Shining my flashlight toward her, I noticed beads of sweat rolling down her forehead. Before I could comment further, Lauren's condition worsened—she began to feel nauseous.

"I don't know what's wrong," she said, her voice strained. "My stomach feels awful. I think I might throw up."

Given Lauren's sudden illness, we decided to relocate to the kitchen, the coolest room in the manor. Upon entering, I noticed something alarming: a long scratch on the left side of Lauren's neck. When she turned, I saw a matching scratch on the right side as well.

"That's probably just from scratching yourself," Bondsy suggested.

"No, I didn't do that," Lauren replied firmly. "And my nails are short; see?" She extended her hand to show us her neatly trimmed fingernails.

So far, the unusual activity seemed to be disproportionately affecting Lauren. This led me to wonder whether she might be particularly sensitive to the manor's spirits—perhaps even empathic. An empath is someone who can sense and absorb the emotions of those around them, experiencing these feelings almost as if they were their own. Given the manor's history of suffering and distress among its patients and staff, it was conceivable that Lauren, possibly sensitive to such energies, was picking up on these residual emotions.

As we huddled in the cool confines of the kitchen, discussing the unsettling possibility that Lauren might be an empath, a sudden realization hit me. At the far end of the Shadow Man hallway, just a short distance from where Lauren's hair had been inexplicably tugged, lay Room Seven. This was not just any room—it had housed a patient who had been known to stand before a mirror, obsessively pulling out her own hair.

Could it be that the phantom responsible for yanking Lauren's hair was none other than the spirit of the woman from Room Seven? Was she somehow envious of Lauren's hair, or perhaps drawn to it in a morbid fascination, compelled to touch or pull it even from beyond the grave?

In my years as a paranormal investigator, the only instance I have experienced of someone's hair being pulled occurred right here at Malvern Manor. The odds of such a phenomenon happening in a corridor adjacent to a room with a history of hair-pulling seemed more than coincidental. If the ghost of the woman from Room Seven still lingered in her room, it's entirely plausible she might extend her presence to the hallway as well, fixated on hair even in death.

We speculated whether the scratches on Lauren's neck and the recent hair-pulling incident could be linked to this restless spirit. With these thoughts swirling in my mind, we decided to step outside for a break, hoping the fresh air might alleviate Lauren's upset stomach and give us a moment to regroup before diving back into the investigation.

After a brief fifteen-minute break, we reconvened in the kitchen, gathering our thoughts before pressing on with the investigation. It was then that something out of the ordinary happened to me—something I hadn't anticipated. Bondsy approached with a worried expression, his usual bravado tempered by genuine concern.

"Larry," he began, his voice low, "the activity seems to be ramping up, especially with Lauren. First, it was just a mild hair pull, then a violent yank, then she got sick, and now she's been scratched. Maybe we should call it a night."

His concern was valid, but something inside me snapped. I felt a surge of anger, sharp and uncharacteristic, rise within me. Without thinking, I lashed out at Bondsy, my voice louder than I intended, my words harsher than I'd ever directed at him before.

"You bring interns along, hoping they'll get scared for good radio," I barked, "but when things actually happen, you want to pack it up and go home? You need to take the paranormal seriously—this isn't just entertainment."

Bondsy, taken aback, tried to calm me down, but I turned on my heel and walked away, still seething. It wasn't like me to lose my temper like that, especially not with Bondsy, who'd been by my side through countless investigations. The outburst left me unsettled, as did the memory of the earlier incident when I'd tumbled down the attic stairs. Could these incidents—the anger, the fall—have anything to do with the allegedly possessed doll I had handled earlier in the night?

Weeks later, I brought up the topic with Bondsy during a conversation. What had unfolded lingered in my mind long after we had wrapped up the investigation. When I asked if he thought the doll might have influenced my sudden quick temper or my fall, he admitted he'd been wondering the same thing.

"I talked with the interns about it," Bondsy said, "and they thought it was strange too. Kate and Lauren both acted out of character during the investigation. Kate's reaction when Lauren's hair got pulled—laughing and crying at the same time—didn't make any sense. And Lauren…she was weirdly calm through everything, even after her hair was yanked, after she got sick, and after the scratches appeared. Normally, she's more anxious, a bit timid even. It's like these things didn't faze her at all."

Bondsy's observations were unsettling, to say the least. The doll—could it have had something to do with the way we all acted that night? We may never know for sure, but the thought lingers, gnawing at the edges of our understanding. It makes you wonder if there's more to this world than we can see, something that can influence us in ways we can't fully comprehend.

Since we hadn't yet explored the attic, I decided to head up there alone while the rest of the group continued investigating downstairs. The attic had an oppressive stillness to it, the kind that makes you question every creak and shadow, but for the hour I spent up there, nothing out of the ordinary happened. No footsteps in the dust, no whispers in the dark—just an unsettling quiet that clung to the air like a shroud.

When I eventually made my way back downstairs, I found the group gathered, a certain calmness having settled over the manor. It was as if the building itself had decided we'd seen enough for one night. We all agreed that it was time to pack up and head home, the adrenaline finally giving way to exhaustion.

As we collected our equipment and prepared to leave, I reflected on the night's events. The investigation had been eventful, to say the least. I don't think the interns had expected much to happen—after all, it was their first real dive into the paranormal. But the night had delivered more than they had bargained for. In fact, it was safe to say that what they experienced went far beyond anything they had anticipated.

For them, the night had been a stark initiation into the unpredictable world of paranormal investigation—a world where the line between fear and fascination is often blurred and where the unknown is always lurking just out of sight, waiting for the right moment to reveal itself. As we left the manor behind, I couldn't shake the feeling that this wouldn't be the last time we'd encounter whatever it was that had made its presence known that night. And I knew, deep down, that the interns would never look at the world in quite the same way again.

EVP Evidence

The following day, I settled in to review the audio and video footage captured during our investigation. Given the personal experiences we'd had, I was hopeful that the recordings would reveal some tangible evidence of paranormal activity. As it turned out, I wasn't disappointed. However, it's worth noting that during the investigation itself, we only heard one sound—a loud noise at 9:49 p.m. All the other EVP (Electronic Voice Phenomena) recordings were captured on our devices but remained inaudible to the team at the time.

9:49 p.m.

The first EVP recorded was the loud crashing sound that both Kate and I heard while walking down the second-floor hallway. The only device that picked up this noise was an audio recorder we'd placed in an empty room on the first floor, a location Kurt had suggested as a hotspot for EVP activity. The sound we captured can only be described as something heavy toppling over, though our subsequent search of the nearby room—and adjacent areas— revealed no obvious source for the commotion.

9:50 p.m.

The next EVP, which I referenced earlier in the chapter, was recorded just a minute later. Following the loud crash, I asked Kate if she had heard it, and she confirmed she had. It was during this exchange that our first-floor recorder captured a whisper, followed by a voice that seemed to be interacting directly with our conversation. After Kate acknowledged the noise, a voice whispered, "No," as if mocking my question. Then immediately after, a male voice said, "Well Kate, you're not going to go prove it!"

What stands out is that whoever—or whatever—was speaking

knew Kate's name. This suggests an awareness of our presence that goes beyond mere coincidence. It's unsettling to consider, but it seems that these phantom voices not only hear us but also recognize us, even as they remain hidden from our view. Or maybe they are closer to us than we realize.

10:46 p.m.

The fourth EVP of the night, recorded in the parlor, is unfortunately only partially clear. A male voice can be heard saying, "Get," followed by what could be either "her" or "them." Regardless of the exact wording, the implication is chilling—it seems to be instructing someone else. This EVP raises more questions than it answers, leaving us to wonder who—or what—this voice was addressing and what its intentions might have been.

10:55 p.m.

The fifth EVP, recorded just before 11:00 p.m., is particularly curious. It, along with a later EVP at 11:40 p.m., seems to be the same voice, though it's clear that whoever is speaking is attempting to disguise it. The voice is high-pitched, reminiscent of the cartoon character Mickey Mouse, and it says, "Hey, did you hear that?" What's strange is that we didn't hear anything at the time of the recording, so it's unclear what the voice was referring to.

11:40 p.m.

The final unexplained voice of the night was captured at 11:40 p.m., and once again, it's the same high-pitched, Mickey Mouse-like voice from the earlier recording. This time, the voice is almost beckoning, asking, "Where are you?"

The peculiar pitch and tone of these voices led me to recall the stories of Grace, the former resident of the manor who was known for disguising her voice. Could it be that what we captured was Grace herself, still playing her games from beyond the grave? Or was it something—or someone—else entirely? Whatever the case, these recordings offer a haunting glimpse into the unseen world that lurks just beyond our perception, waiting for the right moment to make itself known.

Investigation Summary

Malvern Manor was more than worth the long haul from central Illinois. Whatever resides within those walls isn't content to simply observe—it actively seeks interaction with the living. It's a place that beckons, leaving an indelible mark on those who dare to cross its threshold. After our night there, I can say with certainty that it's a place I hope to revisit in the future.

The activity seemed to center on Lauren, and I believe it's because she possesses a sensitivity to spirits—a quality that makes her more attuned to their presence. If a spirit is seeking to communicate, it makes sense that it would gravitate toward someone who can sense it, someone who is open to that otherworldly connection.

But Malvern Manor doesn't discriminate. Whether you're sensitive to spirits or not doesn't seem to matter within its haunted corridors. Even someone like me, who couldn't sense a spirit if it were sitting on my lap, can feel the weight of something ghostly lurking in the shadows. There's an undeniable presence that haunts the manor, something that has chosen, for reasons beyond our understanding, to remain tethered to this world, engaging with those brave enough—or perhaps foolish enough—to encounter it.

Malvern Manor was an experience that will stay with me for a long time, a haunting reminder of the mysteries that lie just beyond the veil of our everyday reality. It's a place that pulls you back, a place that feels unfinished, as if there are still stories to uncover and spirits waiting to make their presence known. Someday, I hope to return, to delve deeper into the enigma that is Malvern Manor, and perhaps to find the answers that still elude me.

WHISPERS ESTATE
MITCHELL, INDIANA

SIX

In the heart of the small town of Mitchell, Indiana, Whispers Estate stands as a beacon for paranormal enthusiasts, its haunted reputation echoing far beyond the town's borders. Ranked as one of the most haunted places in the United States, it draws curious visitors from all over, each hoping to catch a glimpse of the supernatural. The estate's eerie allure has only intensified with the release of the documentary *When the Walls Talk*, which has brought even more attention to this unassuming corner of Indiana.

Intrigued by its reputation, Bondsy and I decided to make the trek ourselves in the fall of 2018. It was a journey born out of curiosity and perhaps a touch of skepticism, fueled by stories that seemed too chilling to ignore. The almost four-and-a-half-hour drive stretched out before us, each mile bringing us closer to what would prove to be a very interesting night.

Whispers Estate was more than just a haunted house; it was a legend in its own right, a place where history and the supernatural intertwined in ways that defied explanation. The house had a story to tell, one that had drawn countless others before us, each leaving with their own tale of unexplained encounters and eerie experiences.

The Haunted History of Whispers Estate

Whispers Estate, nestled at 714 W. Warren Street in the quiet town of Mitchell, Indiana, stands as a towering testament to the lingering presence of the past. This 3,700-square-foot Victorian-style house, with its two stories of aged wood, full basement, and attic, has long been a focal point of both local lore and paranormal intrigue. Believed to have been built in 1894, its earliest known records only hint at the mysteries it holds within its weathered walls.

In 1899, Dr. John Gibbons and his wife, Jessie, acquired the home from its original owners, Dr. George and Sarah White. The Gibbons, known for their compassion, adopted several orphaned children, one of whom was a 10-year-old girl named Rachel. But tragedy soon struck the family. On Christmas Day in 1912, Rachel, in a moment of innocent curiosity, ventured too close to an open flame while peeking at Christmas gifts. The fire that ensued left her badly burned, and she succumbed to her injuries two days later in an upstairs bedroom. Even now, over a century later, visitors to the house claim to hear and see Rachel's ghostly figure darting through the rooms, her presence a haunting reminder of that fateful day.

Rachel was not the only child to meet an untimely end in Whispers Estate. A 10-month-old infant, Elizabeth, also died in the master bedroom under mysterious circumstances. The scent of baby powder still lingers in the air for some, while others report hearing the faint cries of a baby echoing through the room where Elizabeth's life was tragically cut short.

The master bedroom, a place marked by death, also became the final resting place of Jessie Gibbons, who succumbed to double pneumonia. Guests who have dared to sleep in this room have reported waking to the sound of labored breathing and coughing, as if Jessie's last moments were replaying in an endless loop. Some even describe the sensation of an unseen weight pressing down on their chests, as though someone—or something—were sitting on them, suffocating them with a spectral presence. But the most common occurrence is the eerie sound of the closet doorknob jiggling before the door slowly creaks open, seemingly of its own accord.

Dr. Gibbons, a respected physician in Mitchell, had his office on the first floor of the house. Given the era in which he practiced, it's likely that some of his patients met their end within these very walls during his 26 years of service. The house, it seems, has absorbed every moment of sorrow, pain, and death, becoming a repository for restless spirits.

Other deaths have occurred within the estate as well. In the 1950s or '60s, a man died in the upstairs bathroom, while a young boy tragically lost his life after tumbling down the front staircase. Each death has left its mark, adding another layer to the house's already rich tapestry of hauntings.

In 2006, after years of vacancy, the house was purchased by a new owner. As restoration efforts began, so did the strange occurrences. Phantom footsteps echoed through empty rooms, and disembodied voices whispered directly into the ears of unsuspecting guests, as though the spirits were taunting them. It was these eerie voices that gave the house its chilling name: Whispers Estate.

The attic, however, is where the heart of the paranormal activity seems to reside—or so we were told. Guests who have spent the night there report vivid nightmares and the unsettling sensation of someone—or something—trying to force its way into the room. The sound of the doorknob jiggling is a familiar refrain, a harbinger of the eerie events that often follow.

Perhaps the most terrifying entity associated with Whispers Estate is a shadow figure known as "Big Black." This ominous presence, described by psychics as "not of this world," is most commonly encountered in the doctor's room but has been seen in various locations throughout the house. Those who have seen Big Black are left with an overwhelming sense of dread, as though they've glimpsed something that defies all rational explanation.

Reports of earthquake-like tremors, beds and couches shaking violently, and the pervasive scent of pungent cologne, cigar smoke, rancid meat, and dirty medical bandages further add to the house's terrifying reputation. Whispers Estate is not just a house with a haunted history; it is a place where the past refuses to die, where the line between the living and the dead is blurred, and where the echoes of tragedy continue to resonate in the present. Each room holds its own story, a narrative written in the creaks of the floorboards and the whispers in the dark. Here, time seems to stand still, as if the house itself is caught in a perpetual twilight, forever reliving the sorrow and secrets that linger within its walls.

The Investigation

As we pulled up to the house, an unsettling vibe settled over me—something eerily familiar, yet deeply disturbing. Although I am not clairvoyant, it seems that over the years I have developed a sense where I can feel it when something is off or odd about a location. It was the kind of feeling that creeps up on you during a horror movie, when a family arrives at their new home, or someone visits an elderly relative in an old, creaky house. The atmosphere felt thick with anticipation, as if the house itself was holding its breath, daring us to step inside and uncover its secrets.

We were greeted at the door by two female live-in tour guides who gave us a quick overview of the place. Our first stop was the area where Dr. Gibbons once conducted his medical practice. The room was filled with the original remnants of his medical instruments and tools, each one a relic of the doctor's work. An old ledger, meticulously documenting the patients who had sought his care, lay preserved—a silent testament to the life-and-death decisions made within these walls.

As you step into the doctor's quarters, you're immediately enveloped by an air of bygone urgency. The area spans three rooms, each saturated with history and tinged with an unsettling atmosphere. The first room you enter was once the waiting room, now home to an old couch that seems to sag under the weight of years gone by. A glass display case stands nearby, filled with old medicine bottles, yellowed documents, and various medical relics from Dr. Gibbons's practice. Each item a silent witness to the life-

and-death struggles that played out here.

Moving further in, you find yourself in what was once the main office. Now a bedroom, this room feels heavy with the echoes of the past. It's easy to let your mind wander—to imagine the patients nervously awaiting their diagnoses, the doctor making his rounds, the quiet moments when hope or despair would settle in the room like a thick fog. This room was a place where the fine line between life and death was often crossed, and that energy still lingers in the air.

Beyond the former office lies a bathroom, equipped with an old-fashioned bathtub, shower, sink, and toilet. But this seemingly ordinary space once served a far grimmer purpose. It's believed that this was the operating room, where Dr. Gibbons performed surgeries and tended to the gravely ill. The thought of what occurred in this room lingers like a shadow—patients enduring procedures with only rudimentary medical knowledge to guide them, many likely succumbing to their ailments in a time when medicine was as much art as it was science.

Standing in these rooms, it's impossible not to feel the weight of history pressing in. The air is thick with the memories of those who may have taken their last breaths here, victims of an era when the limits of medical understanding often spelled the difference between life and death. The exact number of souls who may have drawn their final breath within these walls remains a mystery. The house, with its history steeped in both healing and tragedy, holds its secrets tightly. Here, in this part of the house, the line between life and death was often precariously thin, and the echoes of those uncertain moments seem to linger in the fabric of the place.

For the most part, the early portion of our investigation was fairly benign, with nothing out of the ordinary occurring. But as the clock inched past midnight, the atmosphere shifted, and we began to experience a series of unsettling personal encounters.

Before delving into the world of the paranormal, I spent over a decade as a private investigator, honing my skills in surveillance techniques. I bring this up because I approach investigating the supernatural in much the same way, using similar methods and equipment. My process is straightforward: I set up audio and video recorders, then quietly position myself in areas of the location known to be hotspots for paranormal activity. If I hear or see anything unusual, I investigate in the hopes of witnessing or experiencing something unexplained. Once the investigation concludes, I review the audio and video evidence, hoping that something beyond the ordinary was captured. Our investigation of Whispers Estate was conducted in this manner.

The one location in the house that our resident tour guides refused to accompany us, was the basement. Their reluctance wasn't just based on eerie tales or a general sense of unease—it stemmed from firsthand experiences that left a lasting impression. They told us about one guest who, while casually descending the old wooden steps, suddenly felt an unseen hand shove them from behind. Though they were fortunate enough to avoid injury, the experience was enough to keep the guest from venturing down there again.

But that wasn't the main reason the basement was off-limits to our guides. One of them recounted an encounter that had shaken her to the core: she claimed to have witnessed a dark, shadowy figure lurking in the basement. What made this figure truly terrifying was not just its ominous presence, but the unsettling detail she described—a set of ram-like horns protruding from its head, giving it the eerie resemblance of a demon straight out of folklore. This chilling vision was enough to convince her that whatever resided in the basement was not something to be messed with, and certainly not something she was willing to face again.

Like many basements, there is only one set of steps leading down, which means that if you see or experience something ghostly, you'll likely have to pass by whatever is there to leave. So, given the stories we'd heard, we, of course, decided to start our investigation in the basement.

The space was relatively small for the size of the house, roughly 200 square feet with a ceiling height of about seven feet. And like most basements, it had an undeniable creepiness, a sense of something unseen lurking just beyond the edges of the dim light.

In the basement, there was a small table set up amid the usual assortment of items you'd expect to find in an old cellar—rusted tools, forgotten boxes, and the lingering scent of dampness. However, tucked away in one corner was a decidedly unusual object: an old ax, its blade dull but still menacing, standing out like a relic of some darker history. Bondsy and I, following our usual routine during investigations, each took seats about five feet apart, settling into the atmosphere that seemed to grow heavier with every passing moment.

We began chatting casually, comparing the eerie vibe of this basement with other locations we'd investigated together. The Sallie House in Atchison, Kansas, also had a basement that exuded a similarly oppressive energy, and the sight of the ax triggered memories of our time at the infamous Villisca Ax Murder House in Iowa, where a similar-looking weapon had been used in one of the most gruesome crimes in American history.

As our conversation drifted between these haunted places, a sudden crackling noise cut through the stillness—a sound like wood being bent to its breaking point, or perhaps something scratching against the microphone of the audio recorder. The noise was sharp, unexpected, and seemed to come from nowhere. I immediately shifted into investigative mode, hoping to coax out whatever presence might be lurking in the shadows. "Make a noise," I called out, my voice steady despite the tension in the air. "Bang something, knock the ax over."

Almost on cue, the crackling sound returned, only this time it was louder, more insistent. What struck us as odd was that Bondsy and I each heard it coming from a different direction. He was certain the noise had come from his right, while I was equally convinced it had come from my right—which was impossible since we were facing each other.

Our equipment—both the video camera and the audio recorder—captured the sound, confirming it wasn't just our imagination playing tricks. The audio recorder, placed on the table between us, picked up the noise much more clearly than the video camera, suggesting that whatever had caused it was close to us, possibly even right beside the table. The realization that something might be lurking so close, yet unseen, sent a shiver down my spine.

Undeterred, I decided to push further. "If you belong here, make that sound again," I challenged, my eyes scanning the dimly lit room. Almost immediately, a different sound emerged from the darkness—a faint, but distinct noise that seemed to come from the space between Bondsy and me. It was as if something had brushed against or moved the audio recorder.

"What the heck?" Bondsy muttered, clearly unnerved.

"Yeah, that was to my right again, bro!" Bondsy added, trying to make sense of the situation.

"I'm hearing it to your left," I declared, my voice tinged with confusion.

The strange sequence of noises had been recorded at exactly 12:21 a.m., according to the timestamp on our video recorder. The disorientation we felt—each of us convinced the sounds were coming from opposite directions—along with the proximity of the noises to the audio recorder, led us to a chilling conclusion: whatever was making these sounds was moving in the space between us, possibly mere inches away, near the table where the recorder sat. The entity, whatever it was, seemed intent on making its presence known, playing with our senses in the most unsettling of ways.

At precisely 12:24 a.m., just moments after the unsettling crackling and scratching noises had died down, our investigation took a turn that would solidify the house's notorious reputation when a clear, unmistakable whisper was captured by our audio recorder. It wasn't just any whisper; it was a voice that seemed to emerge from the very fabric of the house itself, as if the walls were confirming the dark legends that surrounded this place.

It began as I continued our EVP (electronic voice phenomena) session. Electronic voice phenomena, or EVP, for those unfamiliar, is a cornerstone of modern paranormal investigation—a technique that straddles the line between technology and the inexplicable. At its core, EVP involves the use of audio recording devices, often simple digital or analog recorders, to capture voices or sounds that are not audible to the human ear at the time of recording. These anomalous sounds, which may manifest as whispers, distant shouts, or even full sentences, are believed by many to be the voices of spirits or entities attempting to communicate from beyond the veil.

The process of gathering EVPs is deceptively straightforward. Investigators typically begin by selecting a location known for its paranormal activity—a haunted house, an abandoned asylum, a graveyard steeped in local legend. The recording device is then placed in a quiet room or carried by the investigator as they move through the space. After ensuring silence, the investigator might ask a series of questions directed at any potential entities, leaving pauses between each question to allow time for a response. This is where the technique veers into the realm of the unknown. While the human ear may perceive only silence, the recording device might be capturing something altogether different.

The theory behind EVP hinges on the idea that spirits or entities, existing on a different vibrational plane or dimension, can manipulate electronic frequencies to imprint their voices onto recording devices. These entities are believed to use ambient noise, electromagnetic fields, or even the investigator's own voice as a medium to communicate. The result is a recording that, when played back, may reveal sounds or voices that were not present during the initial session.

Skeptics often argue that EVPs are nothing more than auditory pareidolia—the brain's tendency to impose familiar patterns on random noise—or stray radio signals accidentally picked up by the recording device. However, the sheer volume of EVP evidence collected over the years, often under controlled conditions, has given even the most hardened skeptics pause. Some EVPs are startlingly clear, with responses that directly relate to the questions asked, defying easy explanation.

What makes EVP so compelling, and so eerie, is its intimacy. These are not grand, dramatic manifestations but rather quiet, almost covert whispers from the other side. They suggest a world parallel to our own, one that occasionally bleeds through into our reality in the form of a voice, a message, a brief acknowledgment of our presence. For the investigator, the moment they press play and hear a voice responding from the void is both thrilling and profoundly unsettling—a reminder that the line between the living and the dead is far thinner than we might like to believe.

The next question I posed in the oppressive silence of the basement was direct: "If you want us to leave the basement, do something." It was a challenge, an invitation for whatever presence lingered in the dark corners of that space to make itself known. Almost immediately, the air seemed to thicken as the same unsettling noises we had heard earlier—those faint, scratchy sounds and bangs—returned. Bondsy and I exchanged glances, each of us confirming that the other had heard the same thing. But it wasn't what we heard with our ears that made the hairs on the back of our necks stand on end—it was what we didn't hear but which the audio recorder on the table captured with chilling clarity.

As I continued, I added, "Let us hear you move." Almost as if on cue, the recorder picked up an even louder, more intense scratching noise, followed by a distinct banging sound that echoed through the basement. Bondsy's reaction was immediate: "Did you hear that?" he asked, his voice tinged with unease. "Was that you?" I responded, trying to keep my voice steady, but before I could even process what was happening, the recorder captured something that still makes my blood run cold when I listen to it—a loud whisper, seemingly out of nowhere, clearly saying, "No!"

Bondsy, unaware of the whispered response, remarked, "It's to my right." But the voice was unmistakable on the playback, even if we couldn't hear it at the time. It was as if someone—or something—was mocking us, playing a game from the shadows, a voice from the other side taunting us with its presence.

We didn't realize what we had captured until days later, while reviewing the evidence in the safety of daylight. The whisper was faint but unmistakable, its origin a mystery that still haunts me. We couldn't determine if it was a male or female voice, but we knew one thing for certain: it wasn't either of us. So, the question remains—who, or what, was speaking to us in the basement that night? Was it the shadowy figure that had terrified the tour guide, or some other dark presence watching us from the depths of the basement? Or was it something else entirely, a restless spirit bound to that place, eager to make its voice heard from beyond the veil?

Next, we decided to venture up to what is known as the Red Room, so named for its deep, unsettling hue. This room has gained a notorious reputation, not just for its color but for what it is believed to contain. According to reports, a portal—or vortex—runs up through the house from the front parlor, piercing straight through to the attic room on the third floor. The Red Room is said to be the very heart of the house's supernatural energy, a place where the boundaries between our world and the unknown are perilously thin. Guests who dare to spend the night here often report experiences that could rival the most terrifying of nightmares—beds shaking violently, the ominous sound of a doorknob jiggling as if something unseen is trying to enter, and dreams so vivid and horrifying that they blur the line between sleep and waking.

Gwen, one of our tour guides, recounted a particularly disturbing incident that took place in the Red Room, an experience that left her questioning the very nature of reality. It was a quiet morning when

Gwen, feeling brave—or perhaps foolish—decided to test the room's reputation. She planned to spend some time alone in the room, hoping to experience firsthand what so many others had reported. But, like any cautious investigator, she didn't go in completely unprepared. Her friend Jenny stayed downstairs, serving as a lifeline in case anything went wrong. The plan was simple: Jenny would call or text Gwen every thirty minutes to check on her.

The events that followed were nothing short of bizarre. Gwen shut the door to the Red Room behind her at precisely 6:35 a.m. Just three minutes later, at 6:38 a.m., she texted Jenny to let her know she was about to lie down on the bed—fully clothed, without even bothering to get under the blankets. She placed her phone, glasses, and flashlight beside her and settled in. But almost immediately, she heard a strange popping sound, like the noise a piece of new lumber might make as it shifts and settles. It was then that things took a dark turn.

"I put my forearm over my forehead," Gwen explained, "and immediately heard a stern male voice say, 'Wake up!'" She tried to move her arm but found herself paralyzed. At first, she thought it might be her spirit guide—Gwen is a firm believer in such things—but when she mentally responded, "I am not asleep," the situation escalated. Her eyes involuntarily shut, and she felt something cold and clawed grasp her, pulling her downward with a force she couldn't resist.

"I saw two entities," Gwen continued, her voice trembling slightly at the memory. "A male to my right and a female at the foot of the bed. I tried to scream, but no sound came out. I could hear my screams in my head, but physically, I couldn't make a noise." As she was being dragged downward, she realized with chilling clarity that she could see the bed above her, as if she were being pulled into some dark, unseen void. "I felt the place I was being taken to was not a good place," she said. "I knew if it took me down there, it would change my life forever. So, I fought back with everything I had, using only my mind, and finally, I broke free."

When she came to, the entire ordeal had lasted exactly twelve minutes—an eternity in the terrifying grip of whatever had attacked her. Yet Gwen had no recollection of Jenny calling or texting her during that time, despite the plan they had carefully set in place.

When Gwen finally spoke with Jenny, she was shocked to learn that Jenny had called her every thirty minutes as planned, yet Gwen's phone showed no record of receiving or making those calls. It was as if, for those twelve minutes, Gwen had slipped into another realm—one where time and communication operated under entirely different rules. Whether real or imagined, Gwen's story was truly a bizarre and unsettling experience, one that leaves more questions than answers and one that still disturbs her to this day.

Gwen's eerie experience in the Red Room had captivated me, so much so that I felt compelled to spend some time there myself, hoping to encounter the type of strangeness she had described. The plan was simple: Bondsy would stay downstairs in the parlor while I ventured up to the Red Room alone. This room, though small and unassuming, held a reputation that far exceeded its modest appearance. The room itself was fairly compact, with a bed tucked snugly into a narrow alcove, leaving barely an inch of space on either side. A dresser with an old, tarnished mirror stood against the wall, and an ancient electric fan sat in the corner, long past its prime. The walls were painted a deep, almost oppressive shade of red, and the ceiling angled sharply, forming an A-frame that seemed to close in on the room.

I wandered about the space for a few minutes, taking in the heavy atmosphere, and then sat on the edge of the bed, straining to hear the disembodied whispers for which the house was so notorious. But there was nothing—only silence. Determined to replicate Gwen's experience as closely as possible, I decided to lie down on the bed and close my eyes, just as she had done during her terrifying ordeal. For a while, everything remained still. But then, out of nowhere, I began to feel a strange sensation—a dizziness, as though I were losing my equilibrium, despite lying flat. It felt as if the bed itself was moving, shifting beneath me in a way that defied logic. Even when I opened my eyes, the sensation persisted, an unnerving disorientation that lasted several minutes. I couldn't explain it, and the feeling only subsided after I sat up at the foot of the bed, shaking off the lingering unease.

Several days later, when I reviewed the audio from my recorder, I made a chilling discovery. As I walked around the room, after I had my experience on the bed, the recorder had captured two distinct whispers, only a second apart. Both whispers came from the same ethereal voice at 12:59 a.m., their breathy tones just audible over the background noise. Unfortunately, despite listening to the recording numerous times, I still couldn't make out what was being said. But there was no mistaking that something—or someone—had been in the room with me, its presence barely perceptible yet unmistakably real.

Interestingly, while I was up in the attic, trying to shake off the unsettling sensations that had overtaken me, Bondsy remained downstairs in the parlor, alone but for the heavy silence that permeated the house. It was during this time, precisely forty-six minutes later, that our audio recorder—placed strategically on an old piano near the parlor—captured yet another disembodied voice. This time, however, the EVP wasn't a faint whisper like those recorded in the attic and basement. It was clear, distinct, and loud enough to make one wonder how it could have gone unnoticed at the moment.

The voice, as we would later hear during our evidence review, seemed to be saying one of two things: either "Hold up" or the Spanish word for hello, "Hola." The clarity of the recording was startling, yet Bondsy, who was only a few feet away from the piano, had heard nothing when the voice manifested. The fact that such a loud and seemingly direct communication could occur without being physically heard by anyone in the room added another layer of mystery to the EVP phenomena. Was this spirit attempting to communicate with Bondsy, or perhaps with both of us, using the recorder as its medium? Or was it simply a residual echo from a time long past, trapped within the walls of the house, playing out again for reasons we can only hope to someday figure out?

As intriguing and perplexing as the strange noises in the basement and my unsettling experience in the Red Room were, they paled in comparison to what happened next. The most baffling event of the night unfolded while Bondsy and I were investigating the former doctor's area on the first floor. This area, once bustling with patients and procedures, now harbored a lingering sense of unease. The former waiting room, now filled with dusty medical artifacts, led to the doctor's office—now a bedroom—and the old operating room, repurposed as a bathroom complete with a sink, stool, and shower.

Legend has it that many who have dared to explore this area have encountered a shadowy figure known as "Big Black" lurking near the old doctor's office. Some have heard inexplicable jingling sounds, disembodied voices, and even the eerie laughter of children emanating from the operating room. We decided to start our vigil in the bedroom, hoping to catch even a whisper of the strange phenomena that so many before us had reported. The air felt thick, filled with nothing but silence.

After what felt like an eternity of quiet, Bondsy decided to move into the bathroom, where countless surgeries had taken place years ago. The first thing that greeted him was an old bathtub with a closed shower curtain, except for a small twelve-inch gap on the left-hand side. With nowhere else to sit, he opted for the toilet lid, settling in to see if he could stir up any activity.

About ten minutes into his bathroom stakeout, Bondsy's voice broke the silence. "Hey, Larry," he called out, his tone edged with confusion. "I think there's something or someone touching the back of my neck."

"No kidding? For real?" I asked, the seriousness in his voice catching my attention.

"Yeah," he replied. "It feels like I'm being touched all over my neck and shoulder." He described the sensation as similar to walking into a cobweb, that light, tickling touch that sends shivers down your spine. The timing was uncanny, given that just moments earlier, in an attempt to provoke a response, I had loudly invited any spirits in the room to touch one of us. Now, Bondsy was feeling exactly what I had asked for.

Not long after, Bondsy grew visibly uncomfortable, his nerves frayed by the unseen presence. We decided it would be wise to take a short break in the kitchen, where a lone light offered some comfort. After all, being touched by an invisible force in the dark is enough to unnerve even the most seasoned investigator. Our break was brief, as curiosity got the better of me. I wanted to experience the bathroom for myself, to see if whatever had touched Bondsy might try the same with me.

We returned to the doctor's office, where Bondsy settled on the bed while I ventured into the bathroom. Like Bondsy, I chose the toilet lid as my seat, hoping to encounter whatever had made itself known earlier. I hadn't been seated long when I heard it—a faint tapping or scratching sound coming from the wall behind me, eerily similar to the noise we had heard in the basement earlier in the night.

I began explaining what I was hearing to Bondsy, who was filming with a handheld night-vision camera. But as he entered the bathroom, his attention shifted immediately.

"Did you open the shower curtain?" Bondsy asked, his voice tinged with disbelief.

"No, I haven't touched anything," I responded, confused by the question. "I just walked straight over and sat down."

Intrigued, I stood up and moved toward the shower. Sure enough, the curtain, which had been almost entirely closed when we first entered, was now bunched up on the right side, wide open. Neither of us had touched it, so how had it moved? We rewound the footage on the camera to confirm, and there it was—clear as day. The curtain had been fully closed just moments before we left to take our break.

Something—or someone—had opened that curtain while we were out of the room. The validation that came from our video evidence was chilling. We were left to wonder: What kind of entity had the power to manipulate the physical world, in this case a shower curtain, and what was its purpose in moving something so mundane? Both simple questions but still unanswered.

Unfortunately, the opened shower curtain was our last encounter of the night, and with daylight approaching, we ended our investigation. As we packed up our equipment and prepared to leave, the house seemed to settle into a quiet stillness, as if satisfied with the night's events. The energy that had crackled through the air during our investigation seemed to dissipate, leaving behind a discernible silence that felt almost reverent. It was as if the spirits we had sought to communicate with had retreated into the shadows, allowing us a moment to collect our thoughts and reflect on the experiences we had just witnessed.

Pulling away, I cast one last glance at the house, its silhouette looming against the early morning sky. I couldn't shake the feeling that we had only scratched the surface of its secrets. The spirits that resided within its walls had stories to tell, and while we may be leaving, I knew our connection to Whispers Estate was far from over. I resolved to return, to dive deeper into its haunting history, and to unearth the truths that lay hidden in the whispers of the night.

Investigation Summary

Reflecting on our investigation of Whispers Estate, Bondsy and I found ourselves like we have on many of our investigations, grappling with more questions than answers. The strange, disembodied whispers, the unsettling physical sensations, the scratching noises, and the unexplained movement of the shower curtain—each event seemed to hint at a deeper, more elusive truth buried within the walls of this ominous house. Our experiences, captured on audio and video, provided undeniable evidence of something otherworldly at play. Yet, despite the eerie validation these recordings offered, they also underscored the complexity and unpredictability of the paranormal world.

The recorded whispers in the Red Room, the unseen force that touched Bondsy, and the mysterious movement of the shower curtain all seemed connected—part of a larger, invisible presence that inhabited the house. The energy within Whispers Estate was unmistakable, as if the house itself were alive, responding to our presence in ways that defied logic and reason. Each event was a piece of a puzzle that, when assembled, painted a picture of a place where the boundary between the living and the dead was thin, where the past lingered just out of sight, waiting to reveal itself.

But the most daunting realization came with the knowledge that whatever dwelled within the house was aware of us. It interacted with us, played with our senses, and left us questioning not just what we had experienced, but what its intentions were. Was it merely seeking acknowledgment, or was it something darker, more malevolent? The answers, like the spirits themselves, remained just beyond our grasp—elusive and shrouded in mystery.

The experiences we had at Whispers Estate would stay with us long after we walked out the door, not just as memories, but as reminders of the obscure and often unsettling nature of the paranormal. In the end, Whispers Estate had lived up to its reputation—a place where the whispers of the past still echoed through the halls, waiting for those brave enough to listen.

RANDOLPH COUNTY INFIRMARY
WINCHESTER, INDIANA

SEVEN

Randolph County Infirmary

In the summer of 2016, Bondsy, Kasey Schultz—an intern from 99.7 The Mix—and Bondsy's friend, Jennee Jones, joined me on a four-and-a-half-hour, 273-mile journey to one of the Midwest's most infamous haunted locations: the Randolph County Infirmary. Our visit, as you will soon discover, was far from disappointing.

History of the Infirmary

The land on which the Randolph County Infirmary stands was purchased by Winchester County in 1851 with the intention of establishing a poor farm. It was meant to care for those whom society had neglected or cast aside: the mentally and physically challenged, single mothers, the elderly, orphans, and even a few who had run afoul of the law. The residents were expected to maintain the farm, though many were incapable of doing so due to age or disability.

The original wooden structure, built in 1853 to house just 16 residents, was tragically destroyed by fire less than a year after its completion. In 1856, a new two-story brick building replaced it, but poor conditions eventually led to its demolition. By 1898, a new 50,000-square-foot brick building—the very one we were about to investigate—was erected to accommodate the growing patient population. This structure divided its male and female residents into separate wings, with six large wards, several private rooms, a laundry, a kitchen, and separate dining rooms for men and women. The infirmary was part of a sprawling 350-acre property, and just 230 yards northwest of the building lies a graveyard where an estimated 50 souls are buried in unmarked graves.

In 1994, new owners took over and transformed the facility into the Countryside Care Center, which housed 12 residents. By 2009, the Care Center closed its doors, leaving just five remaining residents, who were abruptly relocated. As the story goes, these individuals were told they were going for a ride, only to find themselves at their new residence, never to return to the infirmary.

Winchester County used the building until 2016, when it was sold to its current owners, who now lease it to paranormal enthusiasts for overnight investigations.

Haunted History

Ghost stories seep through every wall of the Randolph County Infirmary, with the sound of slamming doors being one of the most prevalent claims. These echoes often reverberate through the empty halls, sometimes merely as noises, but at other times, people have discovered doors mysteriously closed after hearing these sounds. The slamming and banging are said to be especially common around the building's holding cell, a place reserved for those who became disagreeable or violent.

Footsteps—disembodied and unnerving—are often reported pacing the first- and second-floor halls. But it's not just footsteps that haunt these spaces. Disembodied voices, too, are a frequent occurrence throughout the infirmary. Some claim to hear the voices of small children, their spectral laughter and playful chatter drifting through the building. The first-floor women's wing seems particularly active, with reports of childlike laughter echoing through the area. Yet not all voices are innocent or playful. In certain parts of the building, the agonized screams of unseen entities pierce the silence, sending shivers down the spines of those who hear them.

The Randolph County Infirmary is also known for a few well-established entities, figures who have become almost as famous as the building itself. Among them is Doris M. Addington, a longtime resident and kitchen worker who lived at the infirmary for decades. Born in 1915, Doris was institutionalized at age 11 and remained at the infirmary until her death in 2006. She suffered a nervous breakdown, which led to blindness, and witnessed her father's death from a heart attack in the infirmary's living room. Doris's old bedroom, filled with the porcelain dolls she loved in life, is now a hotspot of paranormal activity. The dolls are said to move on their own, and objects in the kitchen, where Doris once worked, are known to shift without any apparent cause.

Another well-known entity is that of a former judge, notorious for his cruel and harsh punishments. Legend has it that he held hearings in the building's attic and now haunts that very space. His gruff, angry voice has been captured on recordings by past investigators, and many now venture into the attic, hoping to communicate with him. But the judge may not be alone in the attic. A child's tricycle, stored up there, is known to roll around on its own, perhaps pushed by the playful spirits of children who still linger in the building.

Our Investigation

Arriving at our destination, we were mindful of the stories surrounding the Randolph County Infirmary. The accounts of slamming doors, unexplained noises, and the eerie feeling of being watched had been shared by many before us. With its somber history and the lingering presence of those who once called it home, the infirmary held more than just the echoes of the past—it had an atmosphere that seemed to carry the weight of its former residents' experiences.

We hadn't yet ventured into the shadowed hallways or the notorious rooms where spirits were said to remain, but even as we pulled in, there was a sense that this building knew strangers had arrived.

Standing on the threshold of our investigation, we could only speculate about what the night might bring. Would we encounter the phenomena that others had described—the sounds of unseen footsteps, the voices that seemed to come from nowhere? Or perhaps we'd experience that unnerving sensation of being touched by something—or someone—we couldn't see.

One thing was clear as we began: the Randolph County Infirmary was more than just an old building. It was a place deeply connected to its past, and whatever was inside, we were about to find out firsthand. The night ahead would not just be about seeking evidence but about understanding what truly lingers in a place with such a heavy history.

As we unloaded our equipment into the kitchen just off the dining room, it became clear that this would serve as our base of operations for the night. The area, illuminated by working lights and relatively clean compared to the rest of the building, seemed like the most practical choice. It was also a logical spot for taking breaks, offering a brief respite from the intensity of our investigation.

Yet, as we moved in and out, hauling gear and setting up for the long night ahead, I couldn't shake a growing sense of unease. Even though we were the only ones inside the building, the place felt crowded—almost bristling with unseen presences. It was as if the halls and rooms were alive, filled with a silent audience watching our every move.

I'd felt this kind of thing before, usually in graveyards surrounded by dense woods, where the air is thick with the sensation of being observed. The Randolph County Infirmary had a similar vibe. The long, shadowy hallways seemed populated with invisible figures, silently wandering, curious about what these uninvited guests were doing in their domain. It was a feeling I couldn't ignore, and one that lasted throughout the night.

When we first arrived at the Randolph County Infirmary, we were greeted by Adam, the owner of the building, who offered to give us an impromptu tour. As we walked through the dimly lit halls, Adam shared one of his more unsettling experiences in the building—a personal encounter that still lingered with him. One night, while working alone in the basement, he was going about his tasks, not sensing anything out of the ordinary. The place was eerily quiet, but nothing that caused him to be alarmed.

After finishing up in one of the basement rooms, Adam stepped out into the hallway. As he neared the end of the passage, approaching a corridor that branched off to the left, he saw something that stopped him in his tracks—a shadowy figure, the unmistakable silhouette of a man, casually walking past the entrance to the hallway. Initially startled, Adam's first thought was that an intruder had somehow gotten into the building. With a surge of adrenaline, he quickened his pace, determined to catch up with whoever—or whatever—it was.

The infirmary's hallways are long, and by his estimation, he should have been able to see the person within moments. But when he reached the corner and shone his flashlight down the adjoining corridor, there was no one there. He checked the nearby rooms, but they were all empty. The building was as silent as before, yet the presence he had just witnessed felt real and tangible. Standing alone in the darkness of the basement, Adam couldn't shake the realization that he may have encountered something beyond the ordinary—a shadowy entity that simply vanished into thin air.

Adam's story left us intrigued and set the tone for our investigation. Naturally, we decided the basement would be our first stop. This would be the first investigation for the girls with Bondsy and me, and the first paranormal investigation for Kasey overall. However, this wasn't Jennee's first brush with the paranormal—she had been to a few haunted spots around her hometown of Jacksonville, Illinois, a place with its own share of eerie legends. Kasey was visibly nervous, understandably so given that this was her first venture into the world of ghost hunting. Jennee, however, was cool, composed, and perhaps a little too confident. She had a healthy dose of skepticism and seemed more amused than anything by the prospect of encountering spirits.

But as we all soon learned, confidence can evaporate quickly in the face of the unknown. The basement, with its oppressive air, deep shadows, and unexplained sounds, was a different beast altogether. It wouldn't take long before even Jennee's skeptical resolve began to waver.

We first descended into the basement around 7:30 p.m., keen to explore the area where Adam had his unsettling encounter with the shadowy figure. The atmosphere was exactly what you'd expect in a place like the Randolph County Infirmary. The walls were painted white brick, though years of neglect had left the paint peeling, exposing patches of raw brick underneath. It was the kind of setting that screamed "haunted asylum," the air thick with that sense of abandonment, of things left behind. In the dim light, the space seemed to extend into a dark, echoing abyss. Despite the eerie vibe, we didn't witness anything out of the ordinary.

But the feeling of being watched lingered. As we moved through the basement, all of us picked up on strange noises coming from farther down the hallway. It wasn't clear what was making the sound—an echo, a creak of old building materials settling? Maybe, or maybe something more. We exchanged uneasy glances and quizzed the others to see if they, too, heard the noises, but pressed forward, hopeful that our equipment might capture something we couldn't see or sense in the moment.

It wasn't until days later, when I was reviewing the audio recordings from that night, that we realized we had documented something truly chilling. I had placed a digital recorder near the spot where Adam had his encounter, and according to the timestamp, it was at 8:05 p.m. when we captured an EVP. The voice that came through was unmistakably high-pitched, like the voice of a woman.

It wasn't just random noise; there was a cadence to it, as if she were asking a question. To my ear, it sounded like she was asking, "Who are you?"

The voice reminded me of the strange high-pitched tones we'd captured during our investigation at Malvern Manor—eerily similar in both pitch and quality. I compared it to the cartoonish sound of Mickey Mouse: odd, but clearly female. Yet, as convincing as it seemed, I couldn't shake the nagging doubt creeping into my mind.

Was this truly a disembodied voice, or was I falling prey to audio pareidolia, where the human brain interprets random or ambiguous sounds as something recognizable—often as voices or music. It's the auditory counterpart to visual pareidolia, where people see familiar shapes in clouds or faces in inanimate objects. Our brains are hardwired to detect patterns, even when none exist, and in the realm of the paranormal, this tendency can lead to misinterpretation of environmental sounds.

Throughout the night, we spent most of our time exploring different parts of the infirmary, occasionally setting up in various locations in hopes of catching a glimpse or sound of something paranormal. The building had a way of playing tricks on your senses, and though the atmosphere was always heavy, nothing out of the ordinary seemed to happen for long stretches. Jennee, true to form, remained calm and composed. Her experience with other haunted locations gave her a confidence that sometimes bordered on skepticism. But that all changed at exactly 1:11 a.m.

We found ourselves back in the basement—this time in a corridor on the opposite side of where Adam, the building's owner, had encountered the shadow figure. The space had a claustrophobic feel, the air thicker, as if the walls themselves were closing in. Bondsy was in a room nearby, investigating alone. Jennee stood in the doorway of the room, casually observing, while I hung back a few feet behind her, trying to stay focused in the dim light. Kasey, a little farther down the hall, was still getting accustomed to the tension that comes with moments like these on one's first investigation into the supernatural.

Without warning, the quiet was shattered by an unusual sound—a noise that, at first, was hard to place. The closest description I can give is that it resembled someone hiccupping, followed immediately by the distinct sound of two rapid clicks, as if someone had placed their tongue against the roof of their mouth and made a quick, deliberate motion. It was the kind of noise that didn't belong in an empty, lifeless basement. The clicking sound reminded me of the sound we heard and recorded during the math test we conducted at Farrar School in Iowa.

The moment it happened, I instinctively asked if anyone else had heard it. To my relief—and concern—both Jennee and Kasey confirmed that they had. Jennee, who had been standing closest to me, said the noise seemed to come from right behind her, which meant it had originated in the narrow space between us. This wasn't just some distant echo bouncing off the walls or the building settling; it was close—too close.

In that moment, I saw a noticeable shift in Jennee's demeanor.

The steady confidence she had carried throughout the investigation faltered. Her face betrayed a hint of shock, her eyes wide with the realization that whatever had made that sound was right there with us, unseen but present. Up until that point, she had been perfectly comfortable venturing off on her own, walking through rooms solo without hesitation. But after that incident, something had changed. She clung closer to the group, no longer as eager to separate herself from the safety of numbers.

Jennee was visibly shaken, and soon after, she requested a break. We all agreed it was a good idea and decided that I would escort Jennee to the first-floor staging area in the dining room, while Bondsy and Kasey would continue investigating in the basement on their own.

Jennee and I made our way back upstairs to the well-lit dining room area. The sudden shift from the dark, foreboding basement to the relative safety of the dining area was a relief, though the air was still charged with what had just happened.

As we sat down to catch our breath, we began discussing the incident. Jennee, still processing the experience, expressed her disbelief at how clear the noises had been and how startlingly close they had sounded. For someone who had spent the night largely unfazed, the idea that some unseen force may have been lurking mere inches from her was enough to shake her sense of security. It wasn't just the noise that rattled her—it was the feeling of vulnerability that came with it. The realization that whatever was down there with us wasn't just distant, flickering in the shadows; it was right there, watching, waiting.

It was a sobering moment for all of us, reminding us once again that places like Randolph County Infirmary, with their dark histories and countless untold stories, don't always reveal their secrets in grand displays. Sometimes, it's in the small, seemingly insignificant moments—like a click in the dark—that the true weight of the paranormal is felt.

A few days after the investigation, I sat down to review the audio we'd captured that night. It's always a meticulous process—sifting through hours of recordings in the hopes of finding something, anything, that might validate what we thought we'd experienced. As I pressed play on the recorder from the section of the building where we heard the unusual sounds, I wasn't expecting much. But then it happened—just as we'd all remembered. There it was, plain as day: first, the unmistakable sound of a hiccup, followed by those same strange clicking noises.

The playback gave me chills. It was one thing to hear it in the moment, to have that unsettling feeling creep over you, but to have it recorded, to be able to replay it, solidified the experience in a way that made it impossible to dismiss. The sequence of events was crystal clear. You could hear the hiccup, then a few seconds later, the clicks—exactly as Jennee had described, coming from right behind her, and as I had heard, directly in front of me. Whatever had made those sounds had been positioned right between us.

Now that I had the audio evidence in hand, it left us with the most important question: Who, or what, made those curious noises? There were no reasons for the sounds to be there. It wasn't a mechanical malfunction, or some stray noise carried by the wind, because we were in the basement in an area with no windows, so there was no wind to cause the sound. It was clear and unexplainable. We knew Randolph County Infirmary had a reputation for eerie occurrences, but disembodied hiccups! This was truly baffling.

Meanwhile, while Jennee and I were taking our break upstairs, Bondsy and Kasey had their own brush with the unknown—a direct encounter that would leave a lasting impression on both of them. The two had decided to investigate the basement area where the owner had seen the shadow figure, a section of the infirmary that carried its own dark reputation. Unconfirmed reports suggested that a man had taken his life by hanging in this very spot. It was just minutes after Jennee and I had gone upstairs, somewhere around 1:45 a.m., when their experience unfolded.

As Bondsy and Kasey were casually discussing the eerie atmosphere of the basement, something unexpected happened. Out of nowhere, a voice—a female voice—cut through the silence, interrupting their conversation. It wasn't a faint sound or something they could brush off as a trick of the mind. It was clear and distinct enough to make Kasey immediately turn to Bondsy, wide-eyed, asking, "Did you hear that?" Bondsy, just as startled, confirmed he had heard it too. The voice seemed to come from nowhere, but it was unmistakably real.

When Jennee and I rejoined them, they were still visibly shaken by what had happened. They recounted the experience to us in detail, their voices tinged with disbelief. Both had heard the voice at the same time, and it hadn't been a whisper or distant noise—it was as though someone else had suddenly joined their conversation, speaking from an unseen corner of the basement.

In my 25 years of paranormal investigations, I've learned that hearing disembodied voices in real time is a rarity. Usually, voices and sounds are picked up during post-investigation review of audio recordings, often missed by the human ear in the moment. But that night at Randolph County Infirmary, real-time voices were not so rare. Throughout the night, we heard voices echoing down the long, desolate hallways, sometimes sounding as though they were coming from different floors of the building altogether. These weren't subtle sounds either; they were distinct, as if conversations were taking place just out of sight. And it wasn't just our perception—many of these voices were captured on our equipment.

Upon reviewing the audio from the recorder, we had placed in the basement near the area where the alleged hanging had taken place, I found clear evidence of the voice that Bondsy and Kasey had heard. The recorder captured not only their conversation but also the voice that interrupted it. You can hear their discussion flowing normally, and then suddenly, a murmur emerges—followed by a more prominent voice that seems to say either, "You hear that?" or "Where you at?"

The audio file stands as a testament to what we experienced that night. It's one thing to hear something in the moment and doubt your own senses, but to have it confirmed by technology—especially in such a notoriously haunted location—only solidifies the reality of these encounters. The voice on the recorder was more than just a noise; it was a direct interaction, an acknowledgment that we weren't alone in the basement corridors.

Randolph County Infirmary, with its tragic history and lingering spirits, gave us more than we had anticipated. The voices, the noises, the feeling of being watched—they were constant reminders that the past leaves traces, and sometimes those traces are far more than echoes. They speak. And on that night, they spoke to us. It seemed as though the ghosts of Randolph were confused about who we were and were trying to communicate.

Our next, and final, disembodied voices of the night were captured at 2:15 a.m., and again a few minutes later, both recorded in the first-floor room that had once been the longtime home of Doris Addington. Doris, as we had been told, was one of the most well-known former residents of the infirmary, and this room had become something of a focal point for paranormal activity, especially in the women's wing. What unfolded here would leave us wondering whether we had encountered the voice of Doris herself or something else entirely.

Unfortunately, or perhaps fortunately, depending on how you look at it, none of us were present in the room when the EVPs were recorded. My digital recorder, placed in the corner of Doris's room, picked up the first voice—soft and whispery, like someone trying to greet you from another room. The faint "Hi" was unmistakable, and before we had a chance to digest the significance of that, a clear humming sound followed. The humming was almost melodic, undeniably female, and eerily peaceful. Then, as if to contrast the warmth of the humming, a gruffer, more masculine voice chimed in, saying what sounded like "Hello."

The humming, in particular, stood out. Its clarity, and distinctly feminine tone, was compelling evidence. And given that this was the women's wing of the building, it aligned perfectly with the type of residual energy one might expect in a place like Doris's room. But the question lingered—was this humming the voice of Doris in real time, somehow responding to our presence even though we weren't in her room at the time? Or was it something less interactive, something more elusive—a residual haunting? A voice from the past, imprinted on the fabric of the infirmary, playing itself back like an old recording trapped in a loop.

For those unfamiliar with the term, residual hauntings are a concept widely discussed in paranormal circles. Some investigators believe that certain EVPs, like the ones we captured that night, are not the result of conscious spirits trying to communicate but are instead echoes of the past—moments frozen in time, repeating themselves without any awareness of the present. These aren't intelligent hauntings, where entities recognize and react to the living. Instead, residual hauntings are believed to be emotional or traumatic events so powerful that they leave behind a kind of energetic footprint. This energy replays over and over, like a videotape stuck on loop.

In cases like this, you might hear footsteps pacing an empty hallway, or a shadow might glide by in a familiar pattern. The key distinction is that these apparitions, sounds, and voices are not aware of us; they're simply playing back events that happened long ago. In Doris's case, the humming may have been something she often did in life, and that energy—her humming—had become part of the environment, replaying itself long after she was gone.

What makes these types of residual hauntings so compelling is that they offer a glimpse into the past. While they don't communicate or interact, they provide a window into what life—or death—was like for the people who lived within these walls. In that moment, as we reviewed the audio, we couldn't help but wonder: Had we just heard a long-lost fragment of Doris's daily routine, captured and replayed by the very building she once called home? Or was it simply the echo of something much older, left behind in the silence?

The final EVP of the night, captured just minutes after the other recordings in Doris Addington's room, was without a doubt the most striking of the evening. This wasn't just another brief whisper or fleeting hum—what we recorded was the sound of a woman singing. The song lasted for a full eight seconds, a haunting melody that seemed to drift through time. It had that distinct, tinny quality, like something you'd hear playing on an old Victrola or a worn-out record from the 1940s. There was an eerie timelessness to it, as if the voice had been preserved and transported from another era, carried on a whispering wind from the past.

What made this particular EVP even more curious was how

strangely familiar it sounded. I couldn't shake the feeling that I'd heard this voice before. After playing it back several times, it finally hit me—it was almost identical to a recording I had captured a few years earlier at the old Legacy Theatre in Springfield, Illinois. That EVP had also featured the voice of a woman singing, though it lasted much longer than this brief fragment. What was baffling wasn't just the similarity in the tone, but the near-perfect match of the voice itself. The cadence, the timbre—it all sounded like the same woman. Could this be the same spirit, or was it just an uncanny coincidence?

The Legacy Theatre, much like the infirmary, had its own long history of hauntings and strange occurrences, so the possibility of a connection—however unlikely—couldn't be dismissed entirely. Was it possible that this mysterious singer had somehow left her mark on both locations, her voice lingering in these haunted spaces? Or was this simply another example of residual energy—an echo from a past life, repeating itself again and again, detached from time and place?

Either way, this EVP stood out as one of the most compelling pieces of evidence we had gathered. The clarity of the singing, the emotional resonance of the voice, and the way it mirrored the Legacy Theatre recording made it a standout moment in the investigation. Whatever the source of the voice—whether it was Doris Addington, a visitor from another location, or just the remnants of a long-forgotten soul—it left us all questioning just how deep these mysteries truly ran.

As thrilling as it was to capture disembodied voices on our audio equipment, little did we know that the most significant event of the night was still ahead of us. The anticipation that had been building throughout the evening—fueled by distant noises, subtle chills, and the faint sense that we were not alone—was about to be justified in a way none of us could have predicted.

The Randolph County Infirmary had already shown glimpses of its haunted past, giving us the kind of eerie moments every investigator hopes for. But the night was far from over, and the building had more in store for us. What unfolded next would push the boundaries of our understanding, leaving a lasting impact on everyone present. In all my years of investigation, this was about to become one of those nights you never forget—where the inexplicable becomes undeniable, and the line between the living and the dead feels thinner and spookier than ever.

The culmination of our investigation came just after 4:00 a.m. in what had, for the most part, felt like the relatively safe and quiet area of the building—the dining room and adjacent kitchen. Bondsy and Jennee were seated at the dining room table, casually discussing the night's strange occurrences. Kasey had stepped outside for a moment, and I was in the kitchen, just beyond the dining room, packing up our equipment for the long drive home. We were in wind-down mode, going over some of the highlights of the night. The lights were on, giving the room a false sense of security, as if whatever lingered in the building had finally settled down, allowing us to let our guard down. Then, without warning, it happened.

A deafening, thunderous bang echoed through the air—the kind of sound that makes your heart leap into your throat. It wasn't just startling—it was as if the room itself had come alive in that instant. The bang was so loud that I jumped, then froze in place, my back to the dining room, gripping the equipment bag in shock. Bondsy and Jennee jumped too, their conversation abruptly cut off. The noise had come from right behind me, so sudden and violent that I instinctively spun around, half-expecting one of them to be playing a trick. But they weren't.

What I saw instead was the heavy, solid wood door that separated the kitchen from the dining room—now firmly shut. Just moments before, it had been wide open. I stared at the door for what felt like an eternity, processing what had just happened. Finally, I broke the silence. "Did you guys slam the door?" I called out, more confused than anything.

Bondsy, his voice tight with a mix of confusion and fear, shot back, "What the hell? No, we were just sitting here. I was looking right at the door when it slammed shut on its own!"

Still skeptical, I walked over to the door, planning to push it open and join them. I figured maybe it had swung closed on its own somehow, though the force of the slam was impossible to ignore. But when I gave the door a casual shove, it didn't budge at all. I pressed harder, quickly realizing what the problem was. The thick carpet in the dining room was wedged beneath the door, making it nearly impossible to move.

I put my full weight into it and managed to force the door open, inch by inch, until I was finally able to push it back against the wall.

I'm not a small guy—5 feet 11 inches and 200 pounds—and even for me, it took significant effort. The sight that greeted me on the other side was unforgettable. Bondsy and Jennee sat frozen at the table, their eyes wide as silver dollars. The realization was sinking in for all of us.

"I was looking right at the door," Bondsy repeated, his voice a bit shaky. "There's no way it could've slammed like that."

To test it, I yanked the door hard, determined to recreate the event. I pulled with everything I had and tried to slam it shut, but the thick carpet immediately stopped the door's momentum. It wouldn't—couldn't—slam. No matter how hard I tried, the door simply refused to move with the speed and force we had just experienced. I tried several times, and Bondsy gave it a go as well, but the result was always the same. The carpet absorbed all the energy, halting the door before it could even come close to slamming.

The conclusion was undeniable—the door simply couldn't be physically slammed, not under normal circumstances. What had just happened defied the laws of physics and left us all shaking our heads. Whatever force had closed that door, it wasn't anything we could explain, recreate, or see. It was a sobering reminder that the strange events we had experienced that night were far from over, and the building still had its own story to tell.

Investigation Summary

As I reflected on the events that unfolded at the Randolph County Infirmary, it struck me how investigations like these are more than just late-night adrenaline rushes or fleeting moments of fear—they're journeys into the unknown, where the boundaries of reality seem to blur. We came to this forsaken place with our gear, skepticism, and curiosity. We left with more questions than answers, yet somehow a deeper understanding of the mysteries that linger in the shadows of our world.

The Randolph County Infirmary, like so many other places, exists in a kind of liminal space—a place where history, tragedy, and the unknown intersect. It's a place where you can sense the weight of the past pressing in on you, where the echo of long-forgotten voices seems to hang in the air, waiting to be heard. And sometimes—just sometimes—the past reaches out.

There were moments during that night when the paranormal ceased to be an abstract concept and became something visceral, tangible. The disembodied voices, the inexplicable sounds, and, of course, the slamming door—all of these events defied logic and reason. They left us, a team of investigators, standing on the precipice of something we could not explain.

But perhaps that is the nature of these places. They offer glimpses—fleeting and incomplete—into a realm we do not yet fully understand. Whether it was Doris Addington's voice humming softly from beyond the grave or the sudden and unnerving slamming of that heavy door, the infirmary reminded us that the veil between our world and whatever lies beyond is thinner than we might like to admit.

Jennee, once cool and collected, left Randolph County changed, her skepticism now tempered with a newfound respect for what she could not explain. Kasey, in her first investigation, learned that no amount of nerves can prepare you for the chilling reality of coming face-to-face with the unexplainable. Bondsy and I—though we've seen and heard things that would make most people question their sanity—walked away with a reaffirmation of why we continue this work. It's not just about proving the existence of ghosts or collecting evidence of the afterlife; it's about exploring the very fabric of reality and understanding our place within it.

The events at Randolph County will stay with us, like many other investigations from before. They are part of a larger puzzle—one that we may never fully solve but are compelled to keep trying to understand. In the end, that's what drives us to keep returning to these haunted spaces. It's not about the thrill or the fear, but about that enduring question: What happens to us when we're gone?

The Randolph County Infirmary is now just another chapter in our ongoing search for answers. But like every place we investigate, it has left its mark. It's a reminder that, whether through residual energy, intelligent hauntings, or phenomena we haven't even begun to categorize, the past never really leaves us. It's always there, just beyond the corner of our perception, waiting for those who dare pursue it.

ASHMORE ESTATES
ASHMORE, ILLINOIS

EIGHT

Although Ashmore Estates is only a little over ninety minutes from my home in Taylorville, Illinois, it had somehow eluded me over the years. I'd heard the stories—countless tales of haunting activity, some of which were recounted by fellow paranormal investigators whom I trust. It wasn't that I doubted their experiences, but as with any location steeped in paranormal lore, I wanted to experience it firsthand. There's a certain thrill in walking into a place with your own eyes and ears open, not just relying on secondhand accounts.

What better opportunity to finally explore this former poor farm and private psychiatric facility than as the site for the 99.7 The Mix Halloween show? It had the perfect atmosphere—an eerie, isolated property with a long history of hardship and suffering, the kind of place where paranormal activity seems almost inevitable.

In July 2022, with plans in motion for the Halloween show, I joined Bondsy, Josh Roberts from 99.7 The Mix's sister station, 104.5 WFMB, and a mutual friend, Cheryl Ahrens, on a trip to Ashmore Estates. We packed up our equipment, knowing the night ahead could hold anything from subtle EVP captures to full-blown encounters. As we made the short drive to Ashmore, there was an air of anticipation. For me, it was more than just another investigation. It was a chance to finally see if Ashmore Estates lived up to its haunted reputation.

The History and Haunting of Ashmore Estates

In the quiet expanse of east-central Illinois, just beyond the rural community of Ashmore, a foreboding brick building rises against the horizon. Ashmore Estates, now weathered and worn by time, carries with it a long, dark history—one that echoes with the suffering of those who once lived, worked, and died within its walls. Built in 1916, this three-story structure served as the second almshouse for the Coles County Poor Farm, replacing the original building from 1857. For decades, it was home to society's most vulnerable: the destitute, the mentally ill, and those abandoned by family or fate.

From its early days, the poor farm functioned as a final refuge for the forgotten. Between 1916 and 1959, over 250 individuals, referred to as "inmates," passed through its doors. Here, the able-bodied were expected to work the surrounding farmland in exchange for food and shelter, while the sick and elderly endured their days in obscurity. Of those who entered, 32 died within its confines during the first nine years. Their bodies were interred in a county cemetery on the northern end of the property. Eventually, a pauper's cemetery was established nearby, a resting place for between 60 and 100 souls, though the true number remains uncertain—like so much of

Ashmore's history, shrouded in mystery and loss.

In 1959, the poor farm was purchased and transformed into a private care facility by Ashmore Estates, Inc., its mission shifting to serve those with mental illness and disabilities. But this was not an era of hope; rather, it marked a new chapter of hardship. As financial troubles mounted, the facility, like its residents, struggled to survive. By 1986, the doors of Ashmore Estates were closed once more, this time seemingly for good. The building fell into disrepair, abandoned, its empty rooms left to echo with the remnants of lives once lived.

For nearly two decades, the structure sat vacant, a looming relic on the horizon. But in 2006, it was revived—not as a place of care, but as a commercial haunted house. Storm damage in 2013 forced yet another change of ownership, and since 2014, Robin and Norma Terry have worked tirelessly to restore Ashmore Estates, determined to preserve its historical significance while embracing its reputation as a hotbed of paranormal activity.

The estate's tragic past is more than just a story of financial hardship or failed institutions. It is a place where legends thrive, and unexplained phenomena abound. Over the years, eyewitnesses have reported sightings of shadowy figures, eerie thumps, and objects inexplicably moving through the air. Some claim to have been chased up the very stairs where others had met their end, while others speak of being touched by unseen hands—a brush with something from the other side.

Among the spirits said to linger is the ghost of a young girl, possibly Elva Skinner, whose tragic death occurred long before the current structure was even built. In 1880, when the original almshouse stood on the grounds, four-year-old Elva Lowduskey Skinner, born February 28, 1875, died horribly after her clothing caught fire while dressing near a hearth. Her death, one of anguish and innocence lost, is said to leave a lasting imprint—some believe it is her small figure seen wandering near the windows, forever bound to the place where she perished. The deaths of young children catching their clothing ablaze has been a reoccurring story in several haunted places I have investigated over the years, most of which have newspaper documentation to back up the claims.

Then there's Joseph Bloxom, an elderly groundskeeper who met a violent end in 1921. Struck by a train while walking near the tracks, Bloxom somehow managed to drag himself back to the almshouse. Though he was attended to by the superintendent and doctors, his injuries were too severe, and he succumbed the following day. Bloxom's presence, it is said, can still be felt. Some claim to hear footsteps echoing through the halls, as if he is still walking his rounds, never finding rest.

There is also the haunting tale of the doctor who, consumed by his own despair, reportedly threw himself from one of the windows. His apparition, dressed in the formal attire of another era, including a top hat, has been seen gazing out from those very windows—his fate sealed by the same building in which he once practiced. And yet, perhaps the most chilling stories are those of inexplicable encounters: whispers in the dark, a hand gently brushing the back of an unsuspecting visitor, or the sudden drop in temperature that marks the presence of something unseen.

Despite years of investigations and countless stories, the true nature of Ashmore Estates remains elusive. Paranormal investigators flock to the site, drawn by its storied past and the promise of encountering something beyond explanation. Some come armed with high-tech equipment, hoping to capture evidence of the supernatural.

Others arrive out of sheer curiosity, hoping to experience for themselves the strange phenomena that so many have claimed to witness. And while some leave with nothing but a sense of unease, others capture fleeting voices on audio recorders, sudden movements caught on camera, or dramatic shifts in temperature that defy all logic.

The current owners, Robin and Norma Terry, understand the allure of Ashmore Estates. While they have painstakingly restored parts of the building, they have also embraced its reputation as a historic and paranormal landmark, opening its doors to those brave enough to seek answers to the mysteries that linger within. As they work to preserve its legacy, they acknowledge that the building is far more than just an old institution—it is a living relic, one that breathes with the energy of those who came before, and those who never left.

For Ashmore Estates is not just a place where the past lingers— it is a place where it refuses to rest. The spirits of the forgotten, the forsaken, and the lost seem to reach out from the shadows, reminding the living that, here, the line between this world and the next is perilously thin. Those who step through its doors must be prepared to confront not only the ghosts of Ashmore's past, but the chilling realization that, in this place, the dead are never truly gone.

The Investigation

On the night of the investigation, Bondsy, Josh, and Cheryl met me at the local Taylorville Walmart parking lot. It had become the usual spot for Bondsy to meet me when setting out on paranormal excursions that were south of Springfield. We would need to drive separately to Ashmore, with Bondsy riding with Josh and Cheryl with me, as Josh would have to leave at 2:00 a.m. He was filling in for the regular morning show host back at the WFMB studio in Springfield the next morning and would have to leave early to make it back in time.

As we left the familiar streets of Taylorville behind, the conversation between Cheryl and me turned to a particularly spooky investigation we had conducted together several years prior. It was a Civil War–era house that a young couple had purchased with the hope of restoring it to its former glory. The house, however, had other plans.

The couple had begun experiencing odd occurrences soon after moving in. The wife started seeing shadowy figures in the kitchen— dark shapes that seemed to move with purpose but disappeared as soon as they were noticed. Meanwhile, her husband, who had no previous interest in the house's basement, developed an inexplicable obsession with digging up the floor. He claimed to have no idea why, but each time he went down there, his mood shifted. His personality darkened, and there was a palpable tension that grew between them, as if the basement itself was exerting some strange influence over him.

Then there was the third-floor attic. The wife refused to go up there—she wouldn't even speak of it. Something about that space filled her with a dread she couldn't explain. As we delved deeper into the house's history, a sinister past began to reveal itself. Five people had died in the house over the years: three from natural causes, a man who froze to death on the third floor, and another who had tragically taken her own life.

Cheryl and I investigated the house not long after the couple reached out, desperate for answers. The night was interesting, to say the least. Strange knocks resonated from beneath the wooden floor, directly under the sofa where Cheryl and I sat. At first, we thought it might have been settling or some natural creak of an old house. But the knocks had rhythm and intention. Then there was the door—the one that led to the downstairs bedroom. It was heavy and notoriously difficult to open or close. Yet, without warning, it shut on its own. We also recorded several EVPs that we had not heard at the time of the investigation. So, Cheryl had already experienced the strangeness that the paranormal had to offer and was looking forward to finding out what Ashmore Estates would reveal.

As we neared Ashmore Estates, the conversation fell quiet for a moment. I wondered if whatever awaited us inside that crumbling building would surpass even the strange happenings of the Civil

War–era house. Ashmore Estates had a reputation, one that spoke of shadowy figures, unexplained noises, and voices that seemed to drift from rooms long abandoned. We were walking into a place steeped in a century of sorrow and suffering—a place where the past still lingered, waiting for us.

We pulled into the gravel parking lot just before 7:30 p.m., the sun beginning its slow descent behind the flat, windswept plains of Illinois. Ashmore Estates stood before us, its darkened windows and weathered brick walls casting long shadows over the grounds. Robin Terry, the current owner, was waiting for us by the entrance. Robin greeted us with the casual warmth of someone familiar with the macabre, and we soon found ourselves inside the ominous structure, its decaying corridors alive with untold stories.

As we wandered through the empty halls, Robin shared a few accounts of the eerie experiences people had reported since he took ownership of the place. It was clear that he had grown accustomed to the strange and unsettling—perhaps even accepted it as part of Ashmore's legacy. But one story, in particular, cut through the casual air and sent a chill down my spine.

Robin recounted an experience from a local woman who lived just down the road. It had happened around Christmas, a time when her house was festively decorated with lights, ornaments, and holiday cheer. The woman's young daughter had been especially excited about the decorations and seemed proud of the way their house looked from the outside.

"One evening, my daughter came up to me, smiling, and said her little friend loved our Christmas lights—especially the red ones," the woman had told Robin. The statement seemed innocent enough until she asked her daughter who this friend was.

Her daughter pointed toward Ashmore Estates, the dark silhouette of the building visible from their home.

"She told me her friend lived down the road—in that building," the woman explained pointing toward the brick structure, her voice trembling as she recounted the moment. Knowing that Ashmore had been abandoned for years and that its reputation as a haunted site preceded it, the realization hit her with the force of a cold gust of wind. There was no living child at Ashmore Estates.

Robin paused, letting the weight of the story sink in. The thought of a young girl casually interacting with an unseen entity, connected to a place so filled with sorrow and darkness, sent a shiver down my spine. The story, simple yet chilling, encapsulated the essence of Ashmore Estates—a place where the line between the living and the dead blurred effortlessly.

As we continued the tour, Robin's words lingered in the back of my mind. Every creak caused by the wind and every flicker of the dim light seemed to take on a deeper significance. The building wasn't just old—it was alive in a way that defied explanation. And somewhere, in the shadows, a little girl's unseen friend might still be watching the world outside, waiting for someone to notice her.

Once Robin wrapped up the tour of Ashmore Estates, the atmosphere in the building seemed to grow heavier. The shadows stretched longer, and the air carried the weight of untold stories, each corner of the structure holding whispers of the past. We knew it was time to get to work. The building's imposing size, with its three sprawling stories, meant we had to be strategic about where to set up our gear. Paranormal activity had been frequently reported in several key areas, so we focused our attention there.

We began by unpacking our equipment—digital recorders, infrared cameras, and EMF meters. These tools had become second nature to us, essential in documenting the unexplained. We placed them carefully throughout the building. The commissary on the first floor, long abandoned but still echoing with strange energy, was our first target. Then came the room that once housed electroshock therapy sessions, a place where many had likely endured unspeakable pain. On the second floor, the nurse's station area and a closed-in porch, sites of numerous reports of shadowy figures and unexplained noises, were our final setups. With everything in place, the building felt like a trap waiting to be sprung.

Before we officially kicked off the investigation, Bondsy, Josh, and Cheryl decided to make a quick run for provisions—snacks and drinks to get us through what was sure to be a long, eerie night. The nearest convenience store was about eight miles away in Charleston, so I opted to stay behind and keep an eye on the infrared surveillance cameras we'd set up on the second floor.

Alone, I took a seat at the table we'd positioned in the middle of the long, narrow hallway, staring at the video monitor. The glow of the screen was the only light in the corridor, and the silence was thick. The building had its own pulse—every creak and groan from its bones felt alive. For a while, nothing out of the ordinary happened, aside from the occasional blip from the building's security camera, equipped with a motion sensor, near the stairwell. Every now and then, the sensor would be triggered, and the camera would activate, despite the fact that I was the only person in the building. This happened three times in the forty-five minutes that the team was gone.

The camera was positioned at the far end of the hallway, out of my range. There was no rational explanation for the light activating—no one else was in the building. Sitting alone in a place with a history like Ashmore Estates was unnerving enough, but when an unexplained motion sensor light flickers on, it elevates the tension. My senses went on high alert; every nerve in my body buzzed. Something—or someone—was moving out there, and I couldn't see it.

It wasn't until days later, when we reviewed the audio recordings, that the full picture of what had transpired began to take shape. As I listened back to the hours of recordings, I came across a moment that made the hair on the back of my neck stand on end. At exactly 8:45 p.m., during the time I was alone in the building—alone, at least, in terms of the living—a male voice was captured on the recorder. It was clear, unmistakable, and chilling.

The voice simply said, "Hey, fella," as if it were trying to get my attention from beyond the veil. It wasn't distorted or faint; it was as if someone had been standing somewhere in the hallway, speaking directly to me. Yet, in those moments, I had heard nothing.

The voice, although much fainter, was also recorded by our recorder placed in the commissary. Since the voice was recorded at a much higher volume on the second floor, I believe that the disembodied voice more than likely came from the second floor.

Who—or what—was trying to reach out to me that night? Had I been unknowingly in the presence of one of Ashmore's long-forgotten residents, watching me as I sat there, oblivious to the unseen world around me? This clear, disembodied voice was not just an anomaly—it was a reminder of Ashmore Estates' enduring, haunting presence. The building, alive with its dark history, was far from empty.

With my colleagues returning and not knowing we had already captured a mysterious voice in the second-floor hallway, we began our investigation that night in the first-floor commissary. Robin had mentioned this area was particularly active, with multiple paranormal teams recording EVPs and experiencing unusual occurrences. The commissary, separated from the main hallway by a long masonry wall with windows, once served as a hub of activity. Patients gathered here daily for meals, their conversations likely mixing the mundane with the profound. It's a place that holds an imprint of many lives. If only these walls could talk—and perhaps they do—in the form of the EVPs that have been recorded over the years; I'm sure they would tell quite the story.

Around 9:50 p.m., we entered the dark, abandoned commissary. The room was a shadow of its former self, stripped bare of all but a few rusted chairs and remnants of kitchen and dining room furniture.

The space felt heavy with the energy of those who had once passed through. With Robin's suggestion that this was a hotspot for paranormal activity, we decided to settle in, grabbing some of the old chairs and sitting close together in the middle of the room. Our goal was simple: wait, observe, and hopefully hear or record some disembodied voices.

Bondsy, Cheryl, and I took our seats in the dimly lit commissary, while Josh decided to sit just outside in the hallway. The silence was almost oppressive, the air thick with anticipation. Time seemed to stretch as we sat in the dark, our senses heightened to any subtle noise or shift in the atmosphere. We were accustomed to waiting, for long stretches of quiet that often precede the unexpected in paranormal investigations. But that night, at 9:59 PM, the communication came through—not to us directly, but to the handheld video camera that Bondsy was carrying. We didn't hear it at the time, but later, when we reviewed the footage, we discovered it. A clear male voice, out of nowhere, speaking a single word: "Hola." It was distinct and unmistakable—a simple greeting in Spanish, recorded in a place where no living person was speaking.

The oddity of this was not lost on us. A Spanish-speaking entity in this old Illinois building? It seemed strange, yet not entirely unfamiliar. As I listened back to the recording, it hit me: we had encountered something similar before. As you will remember, during our investigation of Whispers Estate in Indiana, Bondsy and I had captured an almost identical EVP. In the dining room of that haunted house, another male voice had clearly said "Hola." The resemblance between the two voices was striking—so much so that I went back to compare them. The tone, the inflection—they were eerily similar, almost as if the same entity had followed us from one location to another.

This wasn't the first time I'd encountered such a phenomenon. In over 25 years of investigating the paranormal, I've recorded voices that seem to reappear, not tied to one specific place but seemingly following us from site to site. It's a rare occurrence but not unheard of. My working theory is that these voices, whether they belong to spirits, entities, or something else altogether, may originate from a parallel dimension—one that occasionally intersects with our own.

I believe that these beings, whoever or whatever they are, know exactly who we are, and know more about us than we realize. They're aware of people like me, those of us who seek to document their existence, who chase after voices in the dark. I don't think these encounters are random. I think they know how to access our world, and in some cases, they want to communicate. And who better to reach out to than someone actively trying to listen for their voices?

It's as if they recognize me, perhaps from previous investigations, and follow, waiting for the opportunity to speak through the equipment we carry. Over the years, I've conducted hundreds of investigations, and the idea that these entities might follow certain investigators, showing up at different locations to communicate, isn't far-fetched. After all, if they can cross dimensions or realms, why couldn't they follow a familiar face, someone they know will be listening?

That night at Ashmore Estates, "Hola" wasn't just a word—it was a bridge between worlds, a reminder that something unseen was watching and waiting, as interested in us as we were in them.

It didn't take long for the night's first unexplained event to shake the stillness of the night. We were still gathered in the darkened room, our conversations reduced to murmurs, primarily focused on listening for any subtle disturbances. The air was thick with anticipation, and every creak of the old building felt like a potential clue. Then, it happened.

Cheryl broke the silence, "Did you hear that?" Her voice was quiet but urgent.

I turned to her, immediately on edge. "What did you hear?" I asked.

She gestured toward the far corner of the room, eyes wide. "Over in the corner," she said.

Before she could elaborate, we all heard it—what sounded like a child's voice mimicking the hoot of an owl. At precisely 10:13 PM, clear as day, we heard, "Whoo, whoo, whoo!" The sound cut through the dark, innocent yet unsettling in its childlike tone. Although it sounded like an owl, there was no mistaking that the sound came from inside the building in the corner. I shined my flashlight, and there was no owl. There was nothing there.

Immediately, Bondsy and I exchanged a glance. "What was that?" we said in unison, the adrenaline rushing through us. Josh, who was sitting just outside the commissary, heard it too. None of us had to explain what we heard; we all knew it was the same—three distinct "whoos" like a child imitating an owl.

We scrambled to check Bondsy's video camera, and sure enough, the camera had captured the strange, childlike voice. It was plain as day on the playback, undeniable. Moments like this remind you why you do this work—it's not the stories you hear from others but the evidence you collect that sends a chill down your spine.

But that wasn't the only voice the commissary would offer up that night. At 2:00 a.m., while we were investigating another part of the building, my audio recorder, still running in the commissary, captured something even more unsettling. This time, it wasn't just an eerie voice but a sequence of events that felt orchestrated, as if something—or someone—was responding directly to us.

Earlier in the night, Cheryl had bravely spent time alone in the pitch-black boiler room, a part of the building that exuded a sense of dread. As we moved through the building, I decided to ask a question out loud, trying to provoke some kind of reaction from whatever spirits lingered in Ashmore Estates.

"If you don't want Cheryl to go back to the boiler room, let us know somehow," I said aloud. "Make some noise or something!"

What came next was a series of sounds that seemed ripped straight from a horror film. First, there was a loud metallic bang— then, right after the bang, came something that raised the hair on the back of my neck the first time I heard the recording of it: a chilling, almost mocking laugh. It wasn't the kind of laugh you'd hear from a person. It was hollow, unnatural, like something straight out of an old scary movie—"Heh, heh, heh."

The laugh was followed by a second metallic bang, just as loud as the first. The sequence was unmistakable: bang, heh-heh-heh, bang.

When we reviewed the recording later, it was clear that the voice wasn't something any of us had heard in the moment. None of us were near the recorder at the time, and no one was anywhere near the commissary when the sounds were captured. Yet the metallic bangs and the unnerving laugh were clear, echoing through the otherwise silent room.

It felt personal, like the building—or something within it—was answering my challenge. Whatever the voice was, it wasn't friendly. It felt mocking, almost like it was toying with us. The way the sounds followed my question, how they seemed to come from just beyond our reach, made me question who—or what—was really in control of the night.

That moment in the commissary left more questions than answers. The voice, the noises—where did they come from? Who, or what, had been lurking in the shadows, playing with us like puppets? I couldn't shake the feeling that something had responded to us, something that knew we were there, watching and listening, but always staying just out of sight.

Like most boiler rooms from a bygone era, the one at Ashmore Estates had an aura that could easily unsettle the hardiest of souls. Its dark corners and rusted pipes gave off the kind of vibe that wouldn't be out of place in a Freddy Krueger film—oppressive, malevolent, like the room itself was watching you. It didn't just feel cold and abandoned; it felt aware, alive in some inexplicable way.

Early in the evening, our team decided to split up. Cheryl and I headed to the Electro Shock room on the second floor, while Bondsy and Josh took the boiler room. The Electro Shock room, with its own haunting history, had a sterile coldness, but it didn't carry the weight of dread that the boiler room did. We had barely settled in for about ten to fifteen minutes when the walkie-talkie crackled.

It was Bondsy. "Hey, were you guys just moving around up there?"

I glanced over at Cheryl, who was still seated in her folding chair. Neither of us had moved. "No, we're both sitting still. Why, what's going on?"

"We just heard movement down here. It sounded like someone was tapping on the concrete floor," Bondsy replied. His voice was steady, but there was an edge to it, like he was trying to process what had just happened.

"Could you tell where it came from?" I asked, intrigued but not yet alarmed.

"Yeah, it came from inside the boiler room. Somewhere between Josh and me—and neither of us was moving when it happened."

I felt a chill run through me. The boiler room had already seemed ominous, but now it felt like it was waking up, responding to our presence. There's always that moment during an investigation when the atmosphere changes, when the building you're investigating stops being just a backdrop and becomes an active participant. This felt like one of those moments.

That wasn't the only strange occurrence in the boiler room that night. When I reviewed the audio recordings a few days later, I discovered something that made my skin crawl. At precisely 12:52 a.m., the recorder we had left inside the boiler room captured a voice. It was loud and clear, like someone standing just a few feet from the mic. The voice said something strange, almost nonsensical: "Cuckoo too!" It didn't make sense, but it was undeniable—a disembodied voice speaking into the silence of the night.

At the time that voice was recorded, we weren't even in the room. In fact, we were down the hall, trying to convince Cheryl to spend a few minutes alone in the boiler room—a task none of us envied. But there was a method to our madness. We believed that one of the spirits haunting Ashmore was a man named Joe Bloxom, a caretaker who had spent a lot of time working in the boiler room. As the story goes, Joe didn't like people, especially women, intruding on what he considered his space.

Joe had lived a hard life, working on the property until May 31, 1921, when he was struck by a passing train while returning to Ashmore on foot. Despite the efforts of the superintendent and doctors, Joe died the following day, succumbing to his injuries. He was buried on the property, and ever since, there have been reports of strange activity in the boiler room—activity many believe is Joe making his displeasure known.

Given Joe's apparent distaste for women in the boiler room, we thought that Cheryl's presence might stir up something. We weren't wrong.

The question that haunted me afterward was whether the voice we recorded was Joe. Was he listening in on our conversation, knowing we were trying to get Cheryl to sit in his boiler room? Or was it some other ghostly entity mocking us with its strange message? Maybe it was just a fragment of residual energy, a leftover piece of some otherworldly puzzle we'd never fully understand. Whatever the case, the voice was real. It had spoken loud and clear, and as we would find out, the boiler room wasn't quite finished entertaining us for the night.

It was after Josh left to head back to Springfield that we were finally able to reluctantly convince Cheryl to spend a short time alone in the boiler room. We hoped her presence might stir up some kind of reaction, whether from the ghost of Joe Bloxom or another lingering spirit. To say Cheryl was apprehensive would be an understatement, but who could blame her? Even in daylight, the boiler room had a menacing atmosphere. The very air inside felt heavy, like it carried the weight of countless secrets—the kinds of secrets that don't rest easy.

Bondsy and I walked her to the room, reassuring her that we would only be moments away if she needed us for anything. As we descended the concrete steps that led into the actual room, the eeriness seemed to intensify. Dust clung to every surface, and the shadows seemed to grow denser the longer you stood there. We found a rusty metal folding chair for Cheryl and wiped it down, clearing away decades of grime before she took her seat.

She sat there, a flashlight in hand, though she only planned to use it if the oppressive darkness became too much to bear. The thin beam of light would be her only bond to the real world. As Bondsy and I headed to our stakeout position—about 40 yards away—listening, watching, and hoping something would break the stillness.

For ten minutes, Bondsy and I waited in the hallway, exchanging hushed words. Every so often, we'd call out to Cheryl, asking if she was okay or if anything unusual was happening. Each time, she replied that nothing seemed out of the ordinary. No sounds. No sudden chills. No whispers from beyond.

From all appearances, her time in the room was uneventful. At least, that's what we thought at the time. But in the world of the paranormal, silence can be deceiving.

It wasn't until I began reviewing the audio recordings from that night that I discovered something—a faint whisper captured on the device we had placed inside the boiler room with Cheryl. The voice was quiet, more of a murmur than anything clearly articulated, but it was undeniably there. A low, breathy sound that seemed to float just beyond the range of comprehension. While we couldn't make out exactly what it said, the tone was unmistakable—like someone, or something, was trying to communicate but just couldn't muster the strength to push through the veil between our world and the netherworld.

The whisper was frustratingly elusive, too faint to decipher, but it was enough to get my attention. There was no one in that room with Cheryl—no living person, that is—who could have made that sound. And yet, the recorder had captured it, as if the room itself had spoken in a voice too faint for human ears to hear.

It wasn't the clear-cut evidence we'd been hoping for, but it was another piece of the puzzle. Another clue that suggested we weren't alone that night, even if we hadn't realized it at the time. Whatever presence lingered there—whether Joe Bloxom or some other restless spirit—it had made itself known, even if only in a barely perceptible whisper.

For Cheryl, the experience may have seemed quiet and uneventful. But for those of us who comb through the details long after the investigation ends, it was one more subtle confirmation that the walls of Ashmore Estates still hold echoes of the past—whispers of the unseen, waiting for someone patient enough to listen.

For me, aside from the rare and awe-inspiring moments of witnessing ghostly apparitions or shadow figures, which I've encountered eight times in the twenty-five years I've been investigating, the most compelling evidence in paranormal investigations comes in the form of EVPs—electronic voice phenomena. There's something undeniably chilling about hearing a voice that shouldn't be there, captured in the dead air of a recorder.

It's that eerie, unmistakable sense that you're not as alone as you think. When I'm not bringing Bondsy or his radio station team along for an investigation, I prefer to work solo. There's a certain purity to it. If I am alone in a remote, forgotten location—whether a dilapidated building or an old rural cemetery—and I capture a voice, there's no doubt. I can rule out the possibility of evidence contamination from another investigator's voice. There's no one to confuse the situation. It's just me and whatever is out there—if anything. Investigating alone sharpens the focus, intensifies the experience, and allows for the cleanest, most credible evidence possible. Despite working solo nearly 80 percent of the time, I've captured over three hundred EVPs—voices and sounds that simply cannot be accounted for.

Ashmore Estates, on this particular night, turned out to be an absolute gold mine for disembodied voices. The audio evidence we gathered provided a rich variety of recordings—some faint, some startlingly clear—that offered more than just anomalies; they presented a narrative, a testament to something lingering in that old asylum. The whispers and voices we documented weren't just fleeting sounds—they were snapshots of an unseen world, fragments of communication from the other side.

EVPs, for all their mystery, are a window into that world, a glimpse into realms we can't fully understand. Capturing one is like pressing a cup against a wall, straining to hear muffled conversations from the other side. They may offer only a few words, a whispered phrase, or even just a sound, but each one feels like a confirmation. A message from the past or another dimension breaks through the thin barrier that separates our world from theirs. The voices we recorded at Ashmore Estates were no different. They were clear enough to be unmistakable, yet mysterious enough to leave us with more questions than answers. Every sound we captured added to the growing pile of evidence that the spirits within Ashmore were anything but silent, and on that night, they seemed particularly eager to speak.

Midway through our investigation, we found ourselves on the second floor, in a room known as Hannah's room. Hannah, a close friend of the building's owner, Robin Terry, had a deep interest in the paranormal and was believed to be sensitive to spirits. Tragically, she passed away in a car accident several years earlier, and Robin dedicated this room to her memory. It held some of Hannah's personal belongings on a shelf and table, along with several photos on the wall.

The team had regrouped after splitting up earlier in the evening. We placed a video camera inside Hannah's room and an audio recorder down the hall, perched on the counter at the former nurses' station. At the time, Bondsy was sitting in a chair to the left of the door, while Josh sat on an old box spring in the center of the room. Cheryl stood near the doorway, a few feet away from Josh, and I was in the back corner, gazing at the mementos that honored Hannah's life.

It was a hot July night, the building's stale air clinging to us. Cheryl, uncomfortable in the heat, turned to Bondsy and asked, "Do you mind if I open one of the windows?"

"Yeah," he replied, gesturing toward the room across the hall. "Open one over there."

What happened next was subtle at first, but unmistakable. After Cheryl returned from opening the window, the air in the room shifted. I had my back to the group, setting a chair between Josh and

Hannah's memorial, when I heard it: a loud, deep breath—or perhaps a sigh—filling the silence.

Immediately, Bondsy reacted. "Was that you?" he asked Cheryl, his voice sharp with curiosity.

"What, the deep breath?" she replied. "It wasn't me."

I chimed in. "I thought it was you," referring to Bondsy.

He shook his head. "It wasn't me."

Then Josh spoke up, pointing in Bondsy's direction. "It came from over there."

Here's where it gets even stranger. Bondsy swore he heard a male voice, clear as day, whispering "Hey!"—but he heard it coming from the hallway just outside the door. Josh also heard "Hey," but he insisted it came from right next to Bondsy. Cheryl, however, only heard a deep breath behind her, and I heard that same breath or sigh near Bondsy, even though I was standing closer to him. It was a bizarre auditory mismatch, almost as if we were all hearing different things in the same moment.

It reminded me of the investigation at Whispers Estate, where Bondsy and I both heard a crackling sound in the basement, but each of us heard it coming from opposite directions. There was something unnerving about how the sound played tricks on us, as though whatever was in the room had the ability to manipulate what we heard.

We decided to review the footage from the infrared camera. The video itself didn't show anything unusual, but the audio was telling. It captured a voice—clear, undeniable. But here's where things got even more perplexing. While Bondsy and Josh both claimed they heard a deep, whispery "Hey," and Cheryl and I had both heard the breath, the camera's audio picked up something else entirely—a high-pitched voice, almost cartoonish, saying "Hey," as if it came from some strange, childlike entity. It was completely at odds with what any of us had experienced.

How is it that Cheryl and I both heard the breath, while Bondsy and Josh heard the voice, and the recorder picked up something altogether different? To this day, none of us have been able to reconcile the discrepancy. It defies logic, as if the paranormal entity was deliberately playing with us, manipulating our senses.

Then, during a broadcast of the station's Halloween show in October 2022, something even more peculiar surfaced. We replayed the audio of the strange voice from Hannah's room, and that's when I heard something we hadn't caught before. It was subtle but unmistakable. While Cheryl was across the hall, just after she had opened the window, I had moved a chair closer to the group. And there it was—a polite, older male voice saying, "Excuse me."

None of us had noticed it during our initial review of the evidence. It was as if an invisible presence had been standing beside me, politely stepping aside as I unknowingly moved into its space. The voice was clear, captured perfectly by the camera's audio.

To this day, we don't know who—or what—was interacting with us on the second floor of Ashmore Estates. None of us had heard the voice during our audio review, but undeniably there it was. It was almost as if an invisible person had been standing near me, and when

I picked up the chair to move it, he—or it—felt they were in my way and politely said, "Excuse me." Whoever or whatever was interacting with us on the second floor remains a mystery. Was it Joe Bloxom, the former caretaker, still making his rounds? Or maybe it was someone or something else altogether, an entity we'll never fully understand. All we know is that the building seemed alive with voices, and not all of them were our own.

The next piece of unexplained evidence surfaced in the Electro-Shock Room on the third floor—a place with a chilling history embedded in its very walls. The room still housed one of the original electro-shock therapy beds, the sole piece of equipment from the building's notorious past. The bed, combined with the few old chairs and a faded sofa in the room, carried an unsettling weight. Knowing what occurred there decades ago—patients strapped down for treatments we now view as archaic and cruel—put everyone on edge. None of us were strangers to the atmosphere of haunted locations, but the Electro-Shock Room had a presence, a residual energy that made its history hard to ignore.

Although we each spent time in the room, none of us experienced anything outright, aside from the usual creaks and thuds you might expect in an old building settling in the night. It wasn't until we later reviewed the recordings that we realized something far stranger had been happening in the room all along.

From 10:06 p.m. to 1:18 a.m., our audio recorder captured a series of bizarre EVPs that defied explanation. The sounds were anything but ordinary, echoing the kind of noise one might expect in a room where patients had been subjected to electro-shock therapy. Yet at the time, none of us heard a thing—even when some of the EVPs were recorded while we were physically in the room.

The first instance occurred between 10:06 and 10:07 p.m. What the recorder picked up was faint, almost imperceptible: an indistinct voice, followed by a rustling sound that seemed too deliberate to be random. When Bondsy heard the clip later, he described it as if someone—or something—was preparing the room, like setting up the bed for a therapy session. It was eerie, to say the least, and set the tone for what followed.

At 10:12 p.m., the recorder picked up a series of hollow bangs, punctuated by a single loud pop. It sounded unnatural, but again, nothing we noticed at the time. Then, at 10:19 p.m., the real strangeness began—a burst of loud static, almost electrical in nature, that lasted for over ten seconds. This wasn't an isolated incident either; the same type of electrical interference was recorded at 11:19 p.m., 11:32 p.m., 11:43 p.m., and again at 11:45 p.m. It was as if something in that room—or in the building—was triggering a disturbance in the audio frequencies. What made it even more curious was that this interference was only recorded in the Electro-Shock Room. None of our other recorders placed throughout the building picked up similar anomalies.

At midnight, Bondsy and Josh ventured back into the Electro-Shock Room. Josh, not one to shy away from a little risk, sat down on the old electro-shock bed while Bondsy took a seat in one of the chairs. They sat in silence, unaware of what was happening just beyond their senses. At exactly 12:06 a.m., the recorder captured another burst of electronic static, but this time there was more. Over the course of 15 seconds, you could hear what sounded like someone nervously tapping their fingers on a counter—or perhaps the long, drawn-out creak of a wooden floor under pressure—followed by what could only be described as the sound of something metallic being dropped into a trash can. When we played back the recording during the station's Halloween show, Josh, who had been sitting on the bed at the time, quipped, "If I'd heard all those sounds while I was up there, I would've gotten off that bed real quick."

The final—and perhaps most perplexing—recording was captured at 1:18 a.m. None of us were on the third floor when it happened. The EVP featured a male voice, though what it said was up for debate—some of us heard "Dusty," while others thought it was asking, "Does he?" The exact words didn't seem as important as the fact that we couldn't explain where the voice came from or what it wanted.

There's another question that still lingers: where did the near-constant static come from? For over three hours, our recorder in the Electro-Shock Room was picking up this persistent, low-level interference. None of the other recorders placed throughout the building—on the first floor, in the basement, or on the second floor—picked up anything similar. Could it be that something in the room itself was causing the static? If it had been some environmental factor, like lightning or electromagnetic interference, surely more than one recorder would have detected it.

It leaves us with the unsettling possibility that what we captured were the residual echoes of past treatments, the lingering energy of something long gone but not entirely at rest. Perhaps the electro-shock sessions left behind more than just scars on the patients who endured them—perhaps they left an imprint on the room itself, replaying in response to our presence. Or maybe something else entirely was stirring within those walls, something that defied our understanding. For now, the answers remain elusive, and the Electro-Shock Room retains its mysteries.

One of the more intriguing rooms at Ashmore Estates is a room Robin Terry, the owner, pointed out as particularly active for capturing EVP evidence. I dubbed it "The Room with a View," for no reason other than its placement near the end of a hallway, with a window overlooking the sprawling fields behind the building. The room itself seemed unremarkable at first glance—just an old vintage sewing machine sitting in one corner. But as with so many places in haunted locations, it wasn't what was there that mattered, but what was there that we couldn't see that counted.

Robin shared a story about a woman who had visited the building. While standing at the window in the room, she saw something extraordinary. In the field outside, she noticed a man—dressed as though from a different time—laboring in the earth, going about his work. Just as quickly as he appeared, he vanished. To her shock, the man had been there one moment and gone the next, another fleeting ghost from Ashmore's long and storied past.

There's a chilling rumor that some former residents of Ashmore

Estates, individuals who lived and died within its walls, may have been buried in that very field. Unmarked graves, forgotten souls—this dark history casts a long shadow over the property, making the "Room with a View" a place charged with quiet intensity.

Though we didn't spend a lot of time in the room during our investigation, I found myself drawn to that window, standing there just as the woman had, staring out into the field. Each time, I hoped to catch a glimpse of something—anything—that would validate her story. But nothing ever came into view. Still, the feeling of being watched, of standing at the edge of a place where history and the present blur, lingered long after I left.

Cheryl, on the other hand, wasn't fond of the room. Each time she stepped inside, she described feeling a sense of unease, an almost perceptible tension. It was as though the room itself carried a weight, a presence that unsettled her in ways she couldn't quite put into words. But despite her reluctance to spend time there, the room lived up to Robin's claims—it proved to be a hotbed of EVP activity.

Our first EVP, captured at 10:29 p.m., was unmistakable. The voice of an elderly woman, clear as day, could be heard saying, "I had to run." The tone was deliberate, almost urgent, as if she were explaining something important, though we were left wondering—what was she running from? That question hung in the air, adding another layer of mystery to the already eerie space.

Minutes later, our recorder picked up something even more unusual—a high-pitched, almost musical sound that lasted for twelve seconds. Whether it was wailing or singing was hard to determine, but whatever it was, it felt deeply human yet otherworldly at the same time. The cadence and pitch didn't fit neatly into any natural category, adding to its strangeness.

This EVP struck a particular chord with me. It immediately reminded me of something we had captured during a previous investigation at Malvern Manor, where we recorded two separate EVPs of a voice that I likened to the high-pitched, cartoonish sound of Mickey Mouse. The voice in "The Room with a View" was identical—an eerie match to what we had captured at Malvern. It wasn't the first time I had recorded this type of voice at different locations, either. Over the years, similar tones had popped up in various places, leaving me with a nagging theory: could it be that spirits sometimes follow investigators from one location to another, recognizing us as conduits, as people who are actively trying to capture their presence? It's a tantalizing thought—that perhaps these voices, these entities, have grown familiar with me and investigators like me, following us from investigation to investigation because they know we are listening and trying to record their voices from the invisible realm.

The final EVP from the room was captured at 12:31 a.m., and this one was perhaps the most startling, even though we didn't hear the voice at the time. The voice, which sounded like that of a young girl—no older than eight or ten—was clear and distinct. What she said was simple, but it sent a chill down my spine when I discovered the voice during audio review: "They are seeing somebody." There was something innocent yet knowing in the way she said it.

Immediately, I wondered—could this be the voice of Elva Skinner, the young girl who tragically died on the property all those years ago? Was she referring to another ghostly entity, or was she asking a question about us, wondering if we were seeing someone?

The "Room with a View" had certainly lived up to its reputation, providing us with a series of EVPs that were not only fascinating but also deeply unsettling. Each voice, each sound, felt like a piece of a larger puzzle, one we may never fully solve. But for now, the voices remain—faint echoes of lives long past, still trying to communicate across time.

In its heyday, the second-floor nurses' station was one of the busiest areas of Ashmore Estates. It's easy to imagine it back then—nurses bustling between rooms, attending to patients, and the sound of footsteps echoing down the hall as they hurried from task to task. Though we didn't spend much time staking out this area during our investigation, we were eager to monitor any potential activity. We placed an audio recorder on the counter, which overlooked a stairwell—a location that would prove to be fruitful in gathering evidence. From this vantage point, the recorder picked up many of the EVPs I've already mentioned, including the eerie "Hey Fella" EVP and the strange, almost guttural "Hoo" sound captured by the recorder in the commissary area.

During the initial walkthrough, Robin shared stories of investigators and visitors hearing voices emanating from the nurses' station. Some had even witnessed shadowy figures roaming about, as though they were still tending to their duties long after the building had been abandoned. And then there was the bell—a small, unassuming object on the counter that, according to Robin, had been known to ring on its own. The idea of a bell inexplicably ringing without human intervention sent a shiver down my spine; it's the kind of phenomenon you want to witness for yourself yet dread at the same time.

The first anomaly captured by the recorder on the nurses' station counter occurred at 9:39 p.m. At the time, Cheryl and I were in the Electro-Shock Room, conducting a brief SB-7 Spirit Box session. I had my reservations about the Spirit Box in the past—skepticism rooted in the randomness of radio frequencies—but over the years, I've experienced too many direct, seemingly intelligent responses to dismiss it entirely. This time was no exception.

In the audio, you can hear Cheryl ask a simple question: "How old are you?" There's no delay, no static-filled gap—just an immediate and crystal-clear response: "Four." For a moment, it was as though time stopped. The voice was unmistakably that of a child, and if you remember from earlier in this chapter, Elva Lowduskey Skinner died at the age of four, just thirteen days shy of her fifth birthday. The sheer relevance of the response was striking.

I've always been cautious not to leap to conclusions, but when a question as specific as "How old are you?" is answered directly by a voice that seems to match the age of one of the building's most well-known spirits, it's hard to chalk it up to coincidence or radio skip. It felt personal—almost too precise. Elva's tragic story seemed to hang in the air, as though she were still there, reaching out across time.

The next EVP recorded at the nurses' station occurred just twenty minutes later, at 9:59 p.m. This time, the audio clip picked up an unmistakable shushing sound, the kind you might hear from someone trying to quiet a room. The sound was clear and deliberate, immediately followed by Josh's voice asking, "Did you hear that?" You can then hear the group murmur in agreement—several members acknowledging they heard it too.

What struck me most about the shushing sound was how familiar it felt. As you'll remember, during our investigation of the YMCA in Granite City, Illinois, our team heard several distinct shushing noises, as though someone was trying to silence us from the shadows. I've experienced this on more than one occasion. At the Legacy Theatre in Springfield, Illinois, I was investigating alone when I distinctly heard someone—or something—shushing me. Not only did I hear the shushing, I recorded it. It's an unnerving experience, to be reprimanded by an unseen presence in a place already filled with tension and mystery.

The frequency of this phenomenon is something I can't quite explain, but it seems that being shushed by invisible entities has become a common occurrence during my investigations. There's something about it that feels almost condescending, as if the spirits are telling us to keep quiet, to not intrude on their space.

Whether it's an intentional act of communication or merely a residual echo from the past, I can't say for certain. But it's a sound that cuts through the silence with a sense of authority—a reminder that we are, after all, the outsiders in these haunted places.

The second-floor nurses' station, with its history of bustling activity, lingering spirits, and mysterious voices, continues to be a focal point of Ashmore Estates. It's a place where the past seems to linger just out of reach, where echoes of a time long gone still resonate in the hallways, waiting for someone to listen. That night, with our recorders quietly capturing the unexplainable, it felt as though we had been allowed a brief glimpse into that world—a world where the living and the dead still share the same space.

Investigation Summary

Ashmore Estates is a place steeped in unsettling history and eerie remnants of the past. From the moment we arrived, the building exuded an aura of dread, its walls seemingly holding onto the suffering and unrest of those who once lived—and perhaps still linger—within. Our investigation began with stories shared by the building's owner, Robin Terry, who recounted numerous strange occurrences experienced by visitors and investigators alike. He pointed out hotspots such as the second-floor nurses' station and the notorious electro-shock room, where chilling activity had been reported.

One of our primary goals was to capture EVPs, and the electro-shock room did not disappoint. Though no one on our team had a direct personal experience while in the room, we recorded a series of unusual EVPs that ranged from rustling noises to electrical interference. One particular recording stood out—a loud pop followed by several seconds of static interference, echoing the room's dark history of shock therapy. At midnight, when Bondsy and Josh sat on the electro-shock bed, electronic disturbances were captured once more, although neither of them heard a thing at the time. The most unnerving moment came at 1:18 a.m., when a disembodied male voice appeared on our recorder, saying something that sounded like "Dusty" or "Does he?"—a question that would remain unanswered.

At the nurses' station, we set up one of our recorders on the counter, hoping to capture the voices that so many had reported hearing. It wasn't long before we did. At 9:39 p.m., while Cheryl and I were conducting an SB-7 Spirit Box session in another room, a clear child's voice came through in response to a simple question. Cheryl had asked, "How old are you?" and immediately the answer came: "Four." This was particularly striking because Elva Lowduskey Skinner, one of the spirits said to haunt Ashmore Estates, died at age four—just days before her fifth birthday.

As the night progressed, the same recorder picked up more unexplained sounds. A distinct shushing noise was captured, eerily reminiscent of shushing noises we'd encountered during investigations at other locations like the Granite City, Illinois, YMCA, which you read about earlier, and the Legacy Theatre in Springfield, Illinois, where I heard and recorded a similar sound. The repetitive nature of this noise across various haunted sites leads me to wonder: are we being silenced by the spirits, or are these just echoes from a world long forgotten?

In one of the more unsettling locations, the "Room with a View," we recorded a series of high-pitched wailing sounds and what seemed to be an elderly woman's voice. The window overlooking the field behind the building carried its own weight of mystery, as Robin shared the story of a woman who once saw a man working in the fields—an apparition from another time. The field itself is rumored to contain unmarked graves, adding to the building's air of sorrow and lost souls.

During the investigation, the EVPs we captured seemed to hint at more than just residual energy. The repeated sounds—whether electrical interference in the electro-shock room or disembodied voices throughout the building—left us shaking our heads in disbelief. Were we hearing the remnants of past trauma, or were these spirits actively reaching out to us? With each recording, the line between the past and the present blurred, leaving us to grapple with the unsettling reality that Ashmore Estates holds onto far more than just memories, revealing its dark secrets one EVP at a time.

EPILOGUE

When we first set out on our investigations, Bondsy and the radio station crew were curious, maybe even a little skeptical. For many, the idea of paranormal encounters was something to be entertained only in fiction, as a quirky topic for late-night radio, or on reality television shows. Yet, over the course of our investigations, something shifted. The unexplained became personal, and the line between skeptic and believer began to blur.

Ashmore Estates, Farrar School, Williamsburg Hill—these places, once just names on a map or locations in a story, became benchmarks for experiences that defied simple explanations. Each site left its mark, not only in the recordings and EVPs we captured but in the minds of those who ventured onto haunted grounds. What started as a curiosity-driven journey turned into an awakening for the team, including the interns who had come along for the ride, never imagining that they'd leave questioning the very nature of reality.

Bondsy, who began as the voice of reason, the skeptic-in-chief, found himself faced with phenomena that reason couldn't explain away. From the math test at Farrar School to eerie, disembodied whispers and cold drafts that accompanied the presence of something unseen, he witnessed firsthand the inexplicable nature of these haunted places. Over time, his incredulous quips gave way to silent contemplation. And he wasn't alone in that transformation. One by one, the team members—each of them grounded in their rational beliefs—had moments that changed them, if not entirely, then enough to admit: there's more to this world than we understand.

The evidence we collected was not just for proof, but for understanding. The EVPs, the shadowy figures, the objects moving on their own—all of these became pieces in a larger puzzle. None of us claimed to have the answers, but we found something undeniable in the questions themselves. We've seen and recorded things that make it impossible to fully dismiss the existence of something beyond the physical.

For me, this journey has never been about convincing anyone. It's about exploration—exploring the boundaries of what we know, challenging the certainties we cling to, and remaining open to possibilities. In every creaking hallway, every chilling EVP, and every shared look of astonishment, we found a deeper connection to the mystery that surrounds us.

In the end, it's not about being a believer or a skeptic. It's about the experience—about walking into the unknown and coming out the other side changed in some small or profound way. Bondsy, the interns, the station staff—they all left these investigations with a little more respect for the world they thought they knew. And while we may never have all the answers, one thing is certain: the unexplained is very real, and it's waiting, just on the other side of belief.

For now, the ghosts of these places remain, not just in the recordings but in our memories. We have learned that the unknown doesn't play by our rules, and perhaps it never will. But the echoes of those nights—the faint, lingering traces of something beyond—remind us that the search is far from over. There will always be another shadow to chase, another voice to capture, another moment when the veil between our world and theirs becomes thin enough to slip through.

Happy hauntings!

Larry Wilson

About the Author

Larry Wilson spent over a decade working as a private investigator before turning his attention to the paranormal. Drawing on his investigative background, he founded **Urban Paranormal Investigations** in central Illinois, where he has explored numerous haunted locations across the Midwest. Known for his careful and methodical approach, Wilson has become a respected figure in the field of paranormal research.

A best-selling author, he has written several books about his experiences, blending true-life accounts with insights into the world of ghost hunting. Larry has appeared as a guest on radio and television programs, as well as at conferences and events where he shares his knowledge and stories of paranormal encounters.

Wilson has also contributed to the production of several paranormal documentaries and can be heard on *The Paranormal Pursuit* podcast, available on major platforms, where he discusses investigations, techniques, and experiences from the field.

He currently lives in Taylorville, Illinois with his wife, Kathy, and their son, Cory.

Photo by Kathy Wilson

Books by Larry Wilson

Chasing Shadows
*Echoes from the Grave (Out of Print)
*Dark Creepy Places (Out of Print)
Where Evil Lurks
Dr. Ugs – A Haunting in Virginia, Illinois
Paranormal Road Trip
Strange Williamsburg Hill
Things That Go Bump in the Night
Grave Encounters
Spooky

www.ingramcontent.com/pod-product-compliance
Lightning Source LLC
Chambersburg PA
CBHW060015100426
42740CB00010B/1498